Reality Conditions

Short Mathematical Fiction

Printed in the United States of America

Current printing (last digit):
10 9 8 7 6 5 4 3 2 1

Reality Conditions

Short Mathematical Fiction

by

Alex Kasman

Published and Distributed by
The Mathematical Association of America

The Spectrum Series of the Mathematical Association of America was so named to reflect its purpose: to publish a broad range of books including biographies, accessible expositions of old or new mathematical ideas, reprints and revisions of excellent out-of-print books, popular works, and other monographs of high interest that will appeal to a broad range of readers, including students and teachers of mathematics, mathematical amateurs, and researchers.

MAA Service Center
P.O. Box 91112
Washington, DC 20090-1112
1-800-331-1MAA   FAX: 1-301-206-9789

Preface

It was not my intention to write mathematical fiction. This all began in the summer of 2000 when I started keeping a website at
`math.cofc.edu/kasman/MATHFICT/default.html`
listing works of fiction that have a strong connection to mathematics.

The list got much larger than I had expected it to and I was seriously impressed with how much there is out there. It seemed to me that fiction with mathematical content was a good way to get people thinking and talking about mathematics without requiring them to make a great effort to learn it in detail. So I wrote a course proposal for an interdisciplinary class in which we would read fiction and discuss its mathematical aspects: What does it say about the role of mathematics in society? What does it say about mathematicians?

Well, I had not been thinking about this for very long before I started noticing gaps in the body of mathematical fiction. There were ideas, stories, and facts that I wished had appeared in some work of fiction but had not. From there, it was not a very big leap to actually writing some stories. So, now you know, each of these stories contains some ideas that I'm trying to get across and is also supposed to be enjoyable to read. (Whether I have achieved these goals is something that I don't know, but which you will after you have read them.)

One danger in stories like these is that they intentionally blur the line between reality and fiction. They include some aspects of reality, perhaps a reference to a real theorem or person, alongside fantasy with no obvious differentiation. It is probably safest for you to assume that everything is fiction, but I hope the notes that appear in an appendix at the end of the book will help to provide some guide as to which parts of the stories are real and which I have made up.

Acknowledgements

I am very grateful to Gerald Alexanderson and his fellow editors at the MAA for their advice and willingness to publish this collection. Several of the stories also benefited from the suggestions and careful reading of Thomas Bullington, a student in my "Mathematics in Fiction" course at the College of Charleston. I gratefully acknowledge assistance from Sabina Hopfer and Pascale Renaud-Grosbras, two visitors to the Mathematical Fiction website with "literary credentials" who were kind enough to share the benefits of their expertise with me. Joe Kelly in the English department at the College also read the manuscript and provided useful comments. And, an inestimable about of credit is owed to my wife, Laura, who supplied not only ideas and suggestions, but more importantly also support and inspiration.

Contents

1

Unreasonable Effectiveness

Amanda Birnbaum began to have second thoughts. Could she really go through with this? Here she was, pacing back and forth in the dry sand on a tiny island she had never even heard of just a few weeks ago.

Though it had a small population of people living in modest houses, the island was unclaimed by any country. As far as any one could tell, it was just so insignificant that no country was interested in it. And, here she was, staring through the wrought iron gate of the only really large house on the island without any idea of who might be residing there. But, she knew she was in the right place. Every major scientific publisher had given her the same description of how to find it. They all had stories about this strange address that, for as long as anyone could remember, had been willing to pay any price to subscribe to all of the top research journals.

Well, even if her crazy ideas were wrong, there was something interesting here. Perhaps she could just find out who it was in this place that felt compelled to keep up on the latest published research! So, she walked up the long path winding between the weeds and wildflowers to the front door and she rang the bell.

Faintly, through the thick wooden door, she could hear the sound of the chimes playing a familiar tune to announce her arrival. She felt a sudden urge to run away, but her curiosity won out and she stayed to see whose feet she could hear scuffling slowly towards the door; to see what sort of person lived here.

The door opened to reveal a short, very ordinary looking middle-aged man with blotchy, dark skin and the whitest hair she had ever seen. When he looked up at her face, he was clearly very surprised and took a step back.

"Oh my," she thought, already comforted by the fact that the man did not look threatening, "he was expecting someone else ... I've frightened him. I'll try to explain. I sure hope he speaks English!"

"Dr. Birnbaum!" he said, now gesturing her in, "I was not expecting to see you here. Please, please, come in. Please, come in."

"You know who I am?!?" Amanda asked as she followed her host around the stone balustrade and into a small sitting room.

"Of course," he answered shyly, indicating a choice of comfortable seating options from a stuffed sofa to a captain's chair. "And why wouldn't I know who you are? Why, everyone is talking about you. Your PhD thesis was published less than one year ago in *Memoirs of the American Mathematical Society* and already has had a profound impact in theoretical biology as well as in your own area of high-dimensional topology. It was brilliant, truly brilliant. I read it myself! Let me get you some tea."

He disappeared through a slender, arched doorway at the back of the room in which she was sitting, giving her a moment to think. If there had been any lingering doubt that she was in the right place, his behavior had eliminated it. She was also glad that he had leapt right into this topic of conversation, since she was never sure how she could bring it up without sounding crazy.

"Yes," she called to the other room, "that is what I had come here to talk to you about." There was no reply, and so she supposed that he could not hear her.

When he returned a few minutes later, he was carrying a silver tea set. Steam flew out of the spout of the teapot as he served. "Ah, *camelia*," he said, sniffing at the fragrant aroma of the tea. "But, I don't understand, why would you come all of this way to talk to me about your work? You can't possibly know who I am!"

"It's not just my research that I want to talk to you about, it is math research in general." Amanda stirred her tea, but did not even lift the cup off of the saucer. She had no interest in the tea just now. "Back when I was an undergraduate, one of my professors had defined mathematics as 'the study of necessary consequences of arbitrary axioms about meaningless things.' My classmates and I, all math majors, did not like this description. It seemed to ignore the *usefulness* of mathematics. After all, math is used in engineering, physics, biology, economics, you know. We had a discussion, about this paradox. 'If mathematics begins with these meaningless abstractions, why is it that it turns out to be so useful in the end?' There were lots of different opinions on the subject."

"Oh yes," he chimed in with a bright smile, "I've heard such debates before many times. 'The Unreasonable Effectiveness of Mathematics'! 'Why is it that results in abstract mathematics, constructed without any thought given to the real world, some time later turn out to be useful after all?'"

"Right! Right. Like non-Euclidean geometry. At the time it was first suggested, it was just a sort of trick. 'Look what we can do if we pretend that parallel lines meet too!' But then, after Riemann, Clifford, Hilbert and Einstein it's no longer a joke, it is a description of the universe we live in, though we never knew it before."

"And," he added, clearly enjoying this conversation, "what about the use of non-commutative rings and imaginary numbers in particle physics?!?"

"Yes," she agreed, "when imaginary numbers were first discussed by mathematicians they were barely even considered to be real mathematics. Now physicists regularly consider quantum wave functions that are complex valued with no qualms."

"And when the physicists first found non-commutativity in their measurements, the supposedly useless theory of abstract algebra was already there, instantaneously transformed into a branch of 'applied' mathematics!" He laughed so hard that she *did* begin to get frightened. He noticed her reaction, and tried to calm himself down, slowly sipping at his tea and trying not to laugh.

"I'm sorry," he said, "I have not had this conversation for quite a long time and ... and I have a special interest in it. Oh, but you had something to say and came such a long way. Perhaps I should just let you talk."

There was an uncomfortable pause while she tried to collect her thoughts and her courage.

"So," she continued, "when my friends and I discussed these ideas in college there were two main viewpoints. Some argued that mathematics allows us to study any structured system, and then since the universe seems to have *some* rules to it, we obviously will be able to use math to say things about it somehow."

"Hmmm," he hummed while nibbling on a sugar cookie.

"And the others all thought that the universe is way beyond our comprehension anyway. According to them, when we have a new mathematical idea, we apply it to the universe because we have nothing better to use."

"Ah," he said swallowing, "as they say: to a man whose only tool is a hammer, everything looks like a nail!" Remembering that he had promised to be silent, he stopped suddenly, 'zipping' his mouth shut with his thumb and forefinger.

"Right. But, I had another idea. It was so crazy, I didn't even mention it to my friends, but I kept it in mind as a sort of joke." She waited for him to ask what the crazy idea was, but he just smiled and looked down at the table, as if he knew she was talking about him.

"My idea," she continued, "was that another good explanation for why 'pure mathematical' research becomes useful some time after its discovery is that the universe itself changes to fit our mathematical discoveries."

"Oh," he said, suddenly blinking rapidly while still smiling. "And why have you come all of this way to talk to me about it?"

"Because, I think you're the one who is doing it."

He nodded slowly, as if admitting it was true. This surprised her, since she had expected a denial. She had expected to be told that she was crazy. It *was* a crazy idea, after all ... wasn't it?

There was another uncomfortable silence.

"So," she said sharply, "it's *true?*"

"Between you and me?" he looked back and forth as if he expected to find people eavesdropping right there in his living room. "Between you and me, it is, and it is a thankless job."

"So, does that make you...you know...are you...the creator of the universe?"

"Ha!" he shouted so loudly that she almost spilled her tea. "If I were one of the creators, you think I'd be here on this crazy little planet in the middle of nowhere? No offense intended. No, no, I've just been doing this here for a few hundred years and after two hundred more I can retire to a nice alternate reality I've been dreaming of."

She was still trying to process all of this. "So, you mean whenever we make a discovery the whole *universe* changes?"

"Well, not quite the *whole* universe, and not quite every discovery. When I find a result that I find especially interesting or entertaining, I find some way to incorporate it into the universe ... but only locally. That's why your cosmologists have been so confused in their theories. In other districts, those with my job may have different tastes in math, different ideas of how 'reality' could be. In fact, it is this diversity of possibilities that the creators enjoy most ... it's why I have a job!" He was very happy to have someone to talk to about this after being silent for so long. A smile of contentment shone on his face and he leaned back in his seat as if he had never been so comfortable in his life.

"But then," she had so many questions, she found it difficult to decide which to ask first, "if ..."

"Wait!" he interupted sitting up straight with a worried expression. "Please, you must tell me how you found me out. I am not supposed to be discovered, you know."

"Well," she said, unable to look him in the eye, "you remember my thesis?"

"Very well!" His eyes lit up in a way she found flattering. He had clearly liked her work. "You noticed that the cohomology of a certain class of high-dimensional manifolds had some bizarre algebraic properties. You called such manifolds 'immunity manifolds' because the behavior reminded you of immune systems. Although you say in the introduction that you are not an expert in biology, the idea motivated your nomenclature throughout the thesis. Some immunity manifolds are healthy, some are not. Some even have auto-immune diseases!"

"Just names I gave them to help me describe and understand the mathematical structure."

"Perhaps, but your scientists were never able to make sense out of the immune system before and there was so much room for rich and beautiful discoveries to grow out of your theories. I just couldn't wait to get to work on it. You must have seen by now how I was able to bring it to play on some questions regarding the improvement of vaccinations and in just a few months some medical researchers working on the disease scleroderma will discover that they can use Serre duality to ..."

"But," she interrupted, "you made a mistake. I mean, *I* made a mistake in proving equation 3.6. The microchimeric subalgebras don't have to be simply connected, and so definition 3.9 just didn't make any sense and ..."

"No," his jaw fell open and he dropped the last little piece of his cookie. "Not equation 3.6! But that was one of my favorite parts. I used that everywhere!"

"Yes, I know. That is how I found you. You see, I caught the mistake just after the *Memoir* was published. It wasn't easy, but I was able to make sure that every copy with the mistake was collected unread and replaced with a corrected version ... every copy except the one that was sent separately by private courier here to your house. And that is how I knew ..."

"Oh my," he said, stirring his tea vigorously. "Oh my, how careless of me! We will have to do something about that, won't we? Yes, something will have to be done about that."

2

Murder, she conjectured

"Look out, Trevor!" Beth screamed, as the tiny car swerved to avoid hitting him. Getting out of his vehicle, the driver surprised the Americans by his size and his attitude. Rather than being angry at Trevor for stepping out in front of the small vehicle, the large man in the woolen cap politely asked "Was anyone hurt?" Then, after a reassuring nod from Trevor and a few angry glances from the drivers of the cars stopped behind his, he was off again down Trumpington Road.

"Trevor, you've got to get this one little fact through your thick skull or you'll never survive this trip," Beth said, only half joking. "This is *England*, so you've got to look *right* before stepping off of the curb, not left!"

"Thanks a lot, but don't you think I know that better than you? Remember, while you've been attending your conference, I've been exploring Cambridge. I've been all over the place, and haven't been killed yet. I just forgot for a moment because we were busy arguing..."

"But that's exactly the point of the argument! You've been exploring Cambridge while I've been sitting in a stuffy lecture hall. Now it should be *my* turn to look around a bit. I don't want to have to hear any more about Zorn's Lemmings!"

"Please, Beth, just one more day? I didn't get to see everything on my list. There's this 'mathematical bridge' at Queen's College, and the Clifton museum, and Isaac Newton's death mask, and Michael Atiyah's house, and just one or two other things I've got to see. I promise that you'll like them too. Okay?"

Though they were approximately the same age, and Trevor was undeniably Beth's equal intellectually, he had a certain innocence about him that was rare among adults. Like a child, he could be simultaneously earnest and naive and selfish in a way that was more endearing than annoying. That was why Beth could hardly ever say "no" to him when there was something that he really wanted. And so, despite her fear that she might be about to spend the most boring day of her life, Beth agreed to go with Trevor to tour the sites of "mathematical interest" that he had not yet found time to see. Trevor, of course, was more optimistic about what the day might hold in store for them, but neither of them imagined that this would be the start of an adventure that they might one day recall as their fondest memory

Trevor and Beth had been dating for almost a year before leaving on the trip but did not know each other as well as this would lead you to expect. If asked, they would say that they loved each other, though they had nothing in common, and that it was their busy work schedule that kept them apart. However, it was clear to all of their friends that this was not quite accurate. In fact, besides the physical coincidence that they were both quite tall and fair, it was a psychological similarity that they shared—an obsession with their own work—that had led a mutual friend to set them up in the first place. It was this same obsessive interest in their professions that kept them from spending very much time together during their yearlong courtship. If they had been in the same line of work, then it might very well have turned out better, but as it happens he was a mathematician and she was a psychologist for the NYPD. Still, when Beth learned that she had to attend a conference in Cambridge during Trevor's summer break, they both thought it would be great to get away together for a vacation.

Now, walking around a small museum dedicated to the nineteenth century Trinity College professor Graeme Clifton, Beth was regretting the decision. This was not the first mathematical site of interest that Trevor had dragged her to, nor the second, nor the third. It seemed to her as if Trevor could not pass a restroom that was once used by a famous British geometer without stopping to pay homage. While touring the Clifton Museum, Trevor was doing his best to explain the significance of this model and that original draft to her, but she just wanted to get out of there.

Trevor was showing her something that *finally* looked as if it might not be completely boring, a geometric model made out of wires that was rigged up to change its shape with the turn crank, when she caught a glimpse of an important word out of the corner of her eye: "Murder." Trevor stopped mid-sentence when he noticed that her attention had been drawn away, and he smiled.

"Yes," he said, "I knew you would like that part ... but I was saving it for last!"

"Can I ... ?" Beth asked, tilting her head in the direction of the yellowed newspaper clipping hanging on the wall.

"Well ..." Trevor said sadly, hoping to make it clear that he still wasn't finished explaining the significance of the different arrangements of straight lines that can be embedded in the Clifton manifold.

"Thanks!"

Trevor walked quickly over to the Plexiglas display, but Beth, who had run over, was already done with the first paragraph of the clipping from the *Times* and started on the second. The front page article from 1870 announced another killing, the fifth strangled woman's body found in London that summer. A plaque beneath the newspaper explained that there were two more killings that summer, and that the murderer was never caught.

"The fifth victim, Heather Clifton, she was his wife?" Beth asked.

"Yes, that's right ..."

"A real unsolved mystery, huh?"

"That's not the only one," he said hopefully, "there's also a mathematical mystery. You see, he only published one more paper after his wife's murder, and it was a bit strange, but quite important really. You see, he made a startling claim in there, the existence of a nice geometric object, a symmetric space, which is the natural space for studying nonlinear differential equations ... like, if you can find the equation you're interested in as a point in this space then the geometry itself somehow tells you what you need to know. Well, anyway, there was no proof of it in the paper and nobody has been able to prove or disprove it since, though we know a lot of things that must be true about this 'space' if it *does* exist. Clifton's Conjecture, as we call it, is so important that many other papers have been written on its consequences, but we don't really know if it is true or not!"

"Kind of like 'Fermat's Last Theorem'?" she asked.

Trevor couldn't believe his ears ... Beth had actually mentioned something mathematical!

"Yes," he said, "a lot like that, only now Wiles has proved Fermat was right—in fact he announced his proof right here in Cambridge at the Newton Institute—while this one is still an open question."

"Let me guess, he was so shaken up by the murder that he just couldn't do the math anymore. Locked himself away. I've seen that sort of reaction before," she said knowingly, "Things are different now, with Prozac and all, but it's never easy to deal with this sort of thing. Especially a random, serial killing like this one."

"Right, kind of *romantic* in a weird kind of way, isn't it? I mean, he needed her so much ..."

"Romantic? No, look, this man was in shock! That isn't romantic it's ..."

This debate was interrupted by a small, old woman whose shoes clicked on the marble floors as she walked towards them waving her arms.

"Wait," said the old woman, whose badge identified her as a museum volunteer. After catching her breath, she continued "Young lady, Clifton's mathematical genius and its connection to Heather is a miraculous story. I cannot let your twenty-first century cynicism destroy every bit of magic left in this love story. Do you know the tale, young man?"

"Well, actually," he said, "I know a lot about his math, but not really much about his life."

"Very well, then," she went on, in an accent ringing with British aristocracy, "listen. Heather Blaine was the daughter of a well-known mathematician."

"No way," Trevor interrupted, "she was Blaine's daughter!? That's really interesting because Blaine's work led to Clifton's in a quite natural way. If Clifton hadn't come along and eclipsed him, I always thought that Blaine would have been the most famous geometer of his day."

"I am flattered that you should say so. You see, Patrick Blaine was my great-grandfather. Heather was my grandaunt. Did you see that picture of her over there, the one in the silver frame? She was absolutely lovely, the most beautiful girl in London. Everyone wanted her, most especially young Graeme Clifton. Yet, Heather was her father's girl, always spending time with him, following him around. She had no time for courtship, and she told Clifton so.

"However, in the month before her 21st birthday, her father died of a burst appendix. They blamed it on his love of Indian food, you know. Well, of course, young Heather was nearly destroyed by this. Clifton, however, had decided that he would win her over and take care of her himself. She still said she would have nothing to do with him, so how could he impress her? With mathematics, of course. He had never had any mathematical training, but just to win Heather's heart, he taught himself advanced geometry and some say he showed himself to be not only her father's equal, but his superior."

"Oh, don't let her offend you!" Trevor called to a portrait of Clifton hanging on the opposite wall. "*Everyone* says you were better!"

"Yes, probably so. Then, to conclude my little story, he eventually won her over with his mathematical ability. But, when she was taken from him, that ability disappeared. To me, it always seemed as if his mathematical talents were a gift of the muses, given to him only so that he could capture Heather's affections."

"Uh-huh," Beth nodded, though she wasn't really listening. She had continued reading the newspaper article and one line especially caught her attention. "It says here that the fifth murder occurred on a Saturday while all four of the earlier ones were Sunday murders."

"Yes," the old woman said, thinking that Beth seemed to have a morbid interest in the murders. "If I remember correctly, the next was on a Sunday also, the very next day!"

"Don't you think that's odd?" asked Beth.

"The whole idea of strangling women to death seems quite odd to me, so the particular day on which the madman chooses to do it makes relatively little difference!"

This was a real conversation stopper. The three of them stood quietly together for a moment each hoping for a way to end the awkwardness, when finally the old woman thought of an especially mysterious aspect of the story that she had never talked about with anyone.

"Actually," she said in a whisper, "I do know one mysterious part of this. I was reading great-grandmother's diary and I noticed a passage about her cousin Michael and how he had *seen* Clifton and Heather together in London that day."

Trevor and Beth waited for this line of thought to continue, but the woman seemed to be finished. "What did he see them doing?" Beth asked finally.

"Oh, nothing, just walking together, but look at the last paragraph of that article you were reading. It says that he was home that day and had *not* accompanied her to London."

Beth and Trevor read it over silently together and saw that it said exactly what she had claimed. "Maybe the paper got that wrong; they always make mistakes like that," Trevor said hopefully.

"Could be," Beth said cautiously, looking straight into the eyes of Clifton in a large portrait hanging besides the exit, "or maybe he had reason to lie."

"Are you suggesting that *Graeme Clifton* was a murderer?" Trevor said later in a Market Street cafe. "Don't you think that's a little unfair considering that he's not even here to defend himself?"

"I'm not saying he was," Beth said, pausing to take one last drag on her cigarette before putting it out in the ash tray, "just that there's something fishy about the whole story. I thought that you'd be *happy* that I'm getting interested in this math stuff. In fact, I'd like for us to stop by the University Library on the way home to look up a few things."

"I'd be happy if you were interested in the math, not investigating a 150-year-old murder!" He signaled to the waitress that they were ready for their check.

"I *am* interested in the math. I am. Let me ask you about some math now ... here you go." Beth passed her empty cup and saucer to the waitress as she slipped the check onto their table.

"Okay, ask away."

"Do you know a lot of mathematicians from the 19th century?"

"Sure, I could name quite a few." Trevor snatched up the check and looked it over to make sure that it was correct and come up with an appropriate tip.

"Any *women*?" Beth asked, reaching into her purse and offering Trevor some coins.

"Women? Sure, off the top of my head I can name Charlotte Angas Scott and Emmy Noether. One American and the other European ... German, I guess. There must be others, too. What are you getting at?"

"Maybe nothing, but I was just thinking it must have been pretty hard to be a woman mathematician in those days."

"I think I see what you're trying to do ... but look I know of two famous mathematical women from about that time, so there's incontrovertible proof that it was definitely possible to be a woman mathematician."

"Trust me, I know from experience that it can be pretty hard to be a woman in a traditionally male occupation."

"Perhaps, but you can't argue with my 'proof by example'." Trevor's chair groaned as it slid across the floor when he stood up. He handed a ten pound note and the check to the waitress who said the word 'lovely' as if she were saying 'thank you' and waved goodbye.

Beth said nothing aloud, but as she followed Trevor out the door she thought to herself "Oh yeah? We'll just see about that."

Trevor had found at least ten books and after briefly perusing them in the aisle, he brought the three most relevant ones to the table where they had agreed to meet. Beth was already there, and from the smile on her face he realized that he was not going to like what she had to say.

"Okay," he asked in a library whisper, "what did you find out?"

"Well, your 'American' mathematician Scott was actually British!"

"Ah! A 19th century *British* woman mathematician! *Exactly* the same as Heather." It was even *better* than he had imagined.

"Well, close enough. She didn't finish college until 1876, six years after Heather Blaine was already dead."

"Yes, well how much could have changed between the days of Heather Clifton's youth and Scott's?"

"Maybe not that much ... "She flipped open one of the books she had found, looking for a particular page that she wanted to show him.

"I don't get it ... you look like you're saying 'I told you so!' but the words coming out of your mouth don't fit with that!"

Beth just smiled and said "Did you know that Scott was an eighth wrangler?"

"What the heck does that mean?!"

The very professorial looking gentleman at the table beside the window was tired of being distracted by Beth and Trevor's conversation and so he let out a loud "Shhh!" They apologized and returned, more quietly, to their conversation.

"I don't know what it means. I was hoping you would know," Beth said sincerely. "It was apparently some sort of serious honor for a math student."

"Sounds very British, but I'm afraid I've never heard of it."

"Anyway, the point is that Scott wasn't even allowed to graduate. They didn't let her get a *degree* in math."

"Because she was a woman?"

"Yes, they didn't let her graduate because she was a woman. So, she went to America. Apparently, we were more accepting of women mathematicians ... or at least female math teachers. Okay, so what did you find out about this Emmy Noether? She must have been a pretty big deal. I saw her name around in some of the books I was looking at."

"Yes, well, she was a very big deal. *Is* a very big deal, in fact. People still refer to Noether's theorem relatively often. It provides a link between dynamics and algebra by showing that ... ow!" She kicked him under the table, her pointy toed boots bruising his left shin. "Oh, sorry. So, I have to admit that Noether had trouble because of her sex too. She was the daughter of a famous mathematician, just like Heather Clifton, and was well known for her research, but even so she was not allowed to teach at the German universities. There was a position, called 'habilitation,' kind of like a postdoc I guess, and women were not allowed to hold this position, which meant they couldn't hold any higher academic position either. Hilbert found a sort of loophole. By saying that she was a 'guest lecturer' in classes that he was officially listed as teaching, she was able to get around it for a time."

"When was this?" Beth inquired. "Was she a contemporary of Heather's?"

"What? No! Don't you know who *Hilbert* was? This was, like, 1914. She was eventually driven out of Germany by the Nazis ... and then she came to the U.S. too."

"I thought you said she was a *nineteenth* century woman mathematician!"

"I guess I wasn't quite right ... about a few things. If it was this hard for them in the 1870's and early 1900's, it would probably have been nearly impossible for a woman to be a mathematician in England in the 1860's."

Beth whispered "Told you so" into Trevor's ear, and for no apparent reason, also kissed him on the cheek.

The next morning, Trevor and Beth used the last of the free meal vouchers that she had been given by the conference organizers to pay for breakfast at the 'buttery bar' of a nearby college. The students with whom they shared their table ate a traditional English breakfast of eggs and meat smothered with baked beans. Neither of the Americans was brave enough to try it, settling instead for a familiar bowl of 'Sultana Bran' and milk.

After breakfast, Trevor and Beth strolled across the bridge at Garret Hostel Lane on their way back to the Clifton Museum. A family of six Japanese tourists disappeared under the bridge and floated out from the other side riding in one of the River Cam's famous punts. The sisters giggled as their adolescent brother, propelling the boat with a wooden pole nearly three times his height, almost slipped and fell into the muddy water.

"But," Trevor said firmly, "that doesn't mean that Heather Clifton *was* a mathematician."

"I guess I don't have enough proof for a die-hard skeptic like you, Trevor. Especially when you still think of Graeme Clifton as a hero, but do you *really* believe that story about him learning math just to impress her and losing the ability as soon as she died?"

"It does sound a bit fishy, now that I think of it." And as they cut across the grassy Backs towards the entrance to the museum, he asked "But, are you suggesting that Graeme Clifton wasn't a mathematician at all?"

Beth held the large wooden door open for him and said "Yes. I guess so. The way it makes the most sense to me is to think that Heather, as a girl, started being interested in math and working on it with her dad. Towards the end of his life, they may have even been publishing results that were really *hers* under his name. Then when he died, she had no outlet for her mathematical ideas ... until she remembered the advances from the sleazy Graeme Clifton ..."

"*Sleazy?!?*" This word echoed, embarrassingly, through the foyer, drawing the attention of two older men who had just finished the tour and were on their way out.

"This is just my theory, okay? Let me roll with it. So, Clifton agrees to marry her and front for her math research. But, after years of doing this, she decides she really wants to tell people that *she's* the mathematician. After all, her work has gotten Clifton a lot of respect, and she knows it is really owed to her. Somehow Clifton finds out, and so when they are in London one day, he kills her and makes it look like part of a serial killing."

"I just *can't* imagine Graeme Clifton as a murderer."

"Think of all that he had to lose!" She did not have to leave this to his imagination. With an outstretched arm she silently referred to the contents of the museum they had just entered. "How do you think he would be remembered if it was known that he was not a mathematician at all, nothing but a sham?"

Trevor had to admit that the theory seemed to hold together; it fit all of the facts they knew so far. But, of course, so did the more accepted story that Clifton was a brilliant mathematician who was emotionally unable to continue his research after he lost his wife through a terrible tragedy. To try to figure out which of these was true, they would need to get more information ... to learn something that would rule one or the other out. And the perfect place to do that was this museum.

"Excuse me," Beth asked at the information desk, "I was wondering whether you could help me find a woman who helped us out here yesterday."

"I'll do my best, young lady. Could you describe her?"

Trevor held his hand out at a height of about two inches below his chin to indicate her size and said "An older woman, 'bout this tall. She claimed to be descended from Patrick Blaine ..."

"*Claimed* to be? Sir, I assure you that Miss Blaine was not lying about this. In fact, she still lives much of the time in the Blaine's old brownstone in London."

"I'm sorry, that's not what I meant ... I just meant ... um, do you know if she's around?"

"She does work this afternoon," the man at the information desk said with a certain finality. "She should be here in about two hours." He turned away from them very deliberately. Beth wondered whether it was just Trevor's choice of words that had annoyed him, their accents, or perhaps he had heard her call Clifton 'sleazy' in the foyer.

"All right," Beth said to the back of the man's head, "We'll just wait in here then." She and Trevor walked through a large glass door into a climate controlled room where visitors can look at some of the less valuable documents and mementos relating to Clifton, each individually wrapped in mylar. They passed the time in there, usually sitting together but occasionally separately, looking for surprises, paradoxes or clues that would shed light on the mystery of Heather's murder. Nothing seemed unusual on the marriage certificate or in any of Clifton's professional correspondence.

"How about this?" Beth asked, knowing that she was reaching. "Did you notice that in all of the pictures of Clifton from *before* the murder, he has a beard and in all of those after he is clean shaven?"

So Trevor looked through all of the pictures again to make sure that she was right, and she seemed to be. But, in the end, he did not find it very interesting.

"I can easily imagine why that might be," he shrugged. "You know, I had a beard for two years in grad school while I dated Daphne because I knew that she went for that look. I hated it, though. Itched like crazy and a pain to keep trimmed! So, you can bet that as soon as she dumped me, I went and shaved it off."

"Yeah," Beth said skeptically, "but don't you think if she'd been *murdered* you would have kept it for ..." She never finished the thought since she was interrupted by the unmistakable sound of the orthopedic shoes of Patrick Blaine's great-granddaughter.

"I *am* surprised to see the two of you here again," she said sincerely. "Mr. Hammond said that you wanted to see me?"

"We did have a few questions, if you don't mind, Miss Blaine. We're still thinking about Heather Blaine's murder. Do you have a moment to spare?"

"Please, do call me Maggie," said Maggie, putting on her volunteer badge, "and I do have to be in the main hall for the next few hours, but I would be happy to answer your questions out there."

"Great!"

They gave Maggie ten minutes or so to get things straightened out as far as the museum went before going to speak to her. First, they gave her a brief summary of Beth's theory. Though she was initially quite surprised by it, Maggie soon seemed to accept the theory as a reasonable possibility. Then, Beth and Trevor began asking questions. It was clear from the very beginning that they did not have any straightforward questions to ask. Since they did not know exactly what they were hoping to learn, the questions were all rather vague.

"So," Trevor began, "can you think of anything that would indicate that Heather was a mathematician herself? Did she ever speak or write about math that you know of?"

"I cannot say that I know all that much about her, so I cannot be certain, but nothing immediately comes to mind. She certainly followed her father around, even when he was working on mathematics, but nobody considered that a sign that she liked mathematics, only that she was entirely devoted to her father." Maggie leaned back in her chair, looked up at the ceiling and bit her lower lip; ever since she was a little girl this is what she did when she was thinking about something very hard. "And I do remember reading in great-grandmother's diaries about how Heather would spend hours alone in her father's library when she visited their home after his death. Again, one could merely imagine that she wanted time to think about her father in one of his favorite spots, but it could also be that she was actually reading his books, working on her research..."

"What about the specifics of the murder itself?" Beth asked, returning the conversation to her particular interests. "I mean, one of the first things that got me thinking about all of this was the anomaly, the strange fact that this is the only murder in the series committed on a Saturday. Was there anything else unusual about Heather's murder as compared to the others?"

"That's certainly not *my* cup of tea, if you know what I mean, but I do know someone who *would* know. There was a young man from Corpus Christi in here just about two years ago who was doing a research project on the murders; I helped him locate some information and items that he needed. *Wilson*, I think his name was. Yes, it was *Webster Wilson* ... how could I forget a name like that?"

"That sounds too good to pass up," Trevor said while Beth nodded in agreement. "Do you think he'd still be in Texas? Could we contact him somehow?"

"*Texas?*" Maggie laughed. "Whatever gave you that idea? No, Wilson's not an American. He's as English as I am, and I would suspect that he's still a student here at Corpus Christi College. Would you like me to try to find his number for you?"

In unison, as if it were a rehearsed routine, Trevor and Beth looked at each other and said "*Oh*, Corpus Christi *College*! Yes, please."

Almost as soon as Maggie handed them a slip of paper with Webster Wilson's current phone number, the couple offered their thanks, promised to return and left again through the grand foyer.

"They waited for you for quite a long time," the man behind the desk said to Maggie, "and then no sooner did you get here than they leave! Strange Yanks."

"Not so strange, I think. They're trying to find the hidden Easter eggs. Remember that game? Running frantically from place to place is part of the fun!"

It's ringing," she told him. The muscles in her cheeks and neck became tighter and tighter as she waited, hardening her generally soft and attractive features. Then, suddenly, her eyes opened wide and her frown brightened to a smile of success. "Hi, am I speaking with Webster Wilson who recently did a project concerning serial killings in London in 1870? Hmmm? No, I'm not from the *History* department ... actually, I'm with the New York City *Police* Department. A colleague of mine, a professor from Queens College in New York and I just need some information about one of those murders and were directed to you as someone who might be able to help us."

Trevor was smiling, partly because he enjoyed squeezing into a little red booth with Beth, partly because of the advertisement that assured him that this was not just a pay phone but a 'BT *Play* Phone', and partly because he wondered what a Cambridge undergraduate would think of "Queens College in New York", but mostly because he liked the idea that he and Beth were working together on a project! He tried to hear Wilson's reply, but it sounded to him like incomprehensible mumbling.

"Yes, I understand that this was only a project for a class, but we are interested in all possible leads. If you could spare just a few minutes of your time we would be very grateful. Perhaps we could take you out for lunch somewhere?"

Twenty minutes later, they were sitting in a booth at a quick curry joint, talking with Wilson over a beer while waiting for their food to arrive. Looking at the patrons of the restaurant, Beth realized that people in England tended to look somewhat disheveled as compared to their American counterparts. Despite her expectation that the 'proper' British would always be dressed impeccably, she saw that mended garments, unshaven cheeks, and dirty hair and clothes seemed to be more acceptable here. In fact, Wilson would easily be mistaken for a homeless person rather than a student if he were spotted in any American college town.

"As I explained on the phone," Wilson said between sips, "I'm not pretending to be an expert on these killings, though I did rather enjoy the project I was working on. The focus of the paper was simply to discuss how the case would have been handled differently today than it was at the time. Evidence collection and analysis, measures to prevent future occurrences and so on ... it was a completely different world then."

Beth was thrilled. "That sounds *perfect*, Mr. Wilson. I would be very interested to hear anything you can remember. For starters, what sort of evidence *did* they have in that case?"

"Barely any, in fact. The fifth murder, a woman from a respected family in London ...

"That was Heather Blaine Clifton," Trevor interjected.

" ... yeah, yeah, that's right! Anyway, her case was the only one where they collected anything that you would clearly label as evidence. I mean, they all had a strangled body and the woman's belongings apparently untouched, but nothing that stood a chance to identify the killer, except in that fifth case."

"That's precisely our interest here ... we are curious about the differences between the fifth murder and the others."

"She put up a fight that was the difference!" Wilson said, leaning back in his chair and sipping on a Czech Budweiser.

"Esteemed guests," the waiter interrupted, "your dishes are ready. Please, enjoy your meal." A prawn vindaloo, flaming red in color as if the spices were hot enough to make the food itself emit a blackbody radiation, was placed in front of each of the men and a masala dosai, too large to fit on the plate, was given to Beth. Trevor's desire to finally try one of the infamous 'vindaloo curries' that he'd heard of from a British friend at home was tempered by the memory that Patrick Blaine's death had been blamed on the same.

Trevor and Wilson each ate a forkful of their lunch, after which Trevor immediately drank half of his beer in a single gulp and Wilson simply said "Smashing" before continuing to describe the evidence in the century old murder mystery. "Interestingly, people at the time seemed to think that it was a *class* thing. I mean, that Heather Blaine Whatever was a woman of class while the other women were not, and so that explained why she struggled with the murderer. From my own experience, I would have thought it would be the other way around ... I mean my own experience with birds, not murders! My guess is that she just caught him off guard, or he had a bad day. Happens to everyone, even psycho-killers."

"Okay, so what kind of evidence was there of the struggle, and from the struggle?" Beth did not need an explanation of why the fifth murder was different; she was more sure than ever that she knew. She just needed evidence to show that she was right.

"You know, just like it says in the police files on the case, all of the other victims were strangled as if the murderer instantly trapped them, killed them and left. No signs of struggle or any evidence that they could find in those days. But the Blaine woman was bruised kind of bad on her arms, the heel of her shoe was busted off almost half a block from where they found the body, and she was clutching a few hairs in her hand."

"Hairs!"

"Right. She had more than a few short, grey hairs tight in her dead hands. Today, of course, that would be perfect evidence—get a pretty good idea of who the murderer was from DNA typing in a police lab—but all it told them was to look for some bloke with grey hair. Not too much of a help when they had no guess as to who it could be."

"Oh, but it is a great help to us!" Beth beamed. "I don't suppose these hairs are still around anywhere?"

"Nah!" Wilson chuckled briefly at her optimism. "As far as I could tell, not very much is left from any of those cases, except some personal belongings from the fifth victim in fact! There's a woman who works at a museum here who is her great-great-granddaughter and she still has the items that were picked up from the police station, the belongings she had on her at the time of the murder."

"You mean Maggie, we know her. Actually, she's the one who directed us to you, but she's just her grandnie ... "

"Forget that, Trevor! Did you look at these belongings?" Beth asked desperately.

"Yeah, I looked at it all. Up in an attic in the house that was her mum's, I guess, they've got her dress, purse, shoes, everything she had on her when she was killed. It's all still in a ratty old box that they got at the police station when they came in to look at the body. But, don't get your hopes up. I didn't see anything there that could identify the killer. Except for the broken heel, it all looked pretty normal."

"Perhaps," Beth sounded less hopeful, "but don't forget, we've got much better techniques these days. We could probably find something, even after all of this time." However, Beth couldn't think of anything they could possibly find that would incriminate Clifton in her murder. Even if they found his DNA on her clothes or belongings, that would be entirely expected in any case since he was her husband. Something in the finality of the last comment was a clue to Wilson that the questioning was over. And so, he decided to make some small talk of his own.

"So, Trevor, I know you're a professor in America. What's your field?"

Trevor was afraid to answer, realizing that Wilson would probably find the truth surprising and bizarre. *I could just lie and say that I'm in criminology or law*, he thought to himself. Nevertheless, he answered sheepishly "I'm a mathematician."

"Maths!" Wilson smiled. "Well then, maybe you would be interested in the victim's personal effects after all. Her purse was full of papers that looked like equations and such. But what does that have to do with anything?"

Trevor and Beth stood on the curb outside the old stone house that had belonged to Patrick Blaine and was now shared by his two greatgranddaughters.

"I'm sure we agreed to meet here at 3 o'clock," Trevor said in a panic. "Could we have the wrong house?"

"It's barely 3:10 now, Trev. Give the lady a chance!"

Just then, they saw her walking up the street. Still a block away, she waved to the couple waiting for her at her weekend home. When she was in shouting distance she called "I think I'm getting too old for this!", and when she was closer "I guess old ladies are supposed to pick a place to live and just sit there, but I'd never be able to decide between here and Cambridge. So, I'm always traveling back and forth. Still, Elizabeth will be very surprised to see me here on a Thursday."

"I don't think she's home. We tried knocking." Trevor looked up at the intricate stonework, the crystal chandelier visible through the transom, and the three-stories of beautiful brownstone in central London, and tried to imagine by what order of magnitude the price of this house exceeded his entire net worth. "Is the house in Cambridge like this one?" he asked, realizing that the question was not entirely polite.

"No," Maggie unlocked the door with two keys, and swung it open to reveal a marble floored hallway, "the one in Cambridge *is* a bit nicer. Won't you come in? I hope you don't mind my being unorthodox, but I suggest that we skip the formalities and head straight for the withdrawing room. Can I offer you tea, or coffee if you prefer?"

Trevor felt like yelling, "No, just show me those papers!" but said nothing while Beth politely said "Tea would be great ... we both like tea."

Maggie walked them through a grandiose ballroom with a high, carved plaster ceiling and beautiful chandelier to a small, comfortably furnished room. While waiting for the tea to be served, Trevor and Beth sat on the sofa and considered the oil painting over the fireplace. "I know that's not Clifton and Heather," Beth said, "because I remember what Clifton looked like from the photos at the museum."

"That's Patrick Blaine," Trevor agreed, "and I guess his wife. I don't even know her name, but I recognize Blaine for sure."

Maggie entered confidently carrying the tea service and biscuits on a tray. "She's Margaret Blaine, just like me." As she poured, she said "You look like you've solved your mystery. Was it something that Webster Wilson told you?"

"We're not quite done with the mystery yet, but at least we're getting somewhere." Beth dunked her biscuit into her tea, immediately regretting it as it broke apart into a dozen floating crumbs. "Can I ask you something?"

"Certainly, that's why I'm here."

"What color was Graeme Clifton's hair?"

"His *hair*? I suppose there is no color picture of him at the museum ... you're right. We'll have to do something about that."

"Color picture?" Trevor said skeptically. "There can't be any color pictures of Graeme Clifton; he died in 1898!"

"Not color photographs," Maggie corrected, "but we have color paintings. In fact, there is one upstairs of Clifton and Heather shortly after their wedding. His hair is a beautiful red, flame red."

This was not good news for Beth, but there was still a possibility. "What about later, like when Heather was killed. Had he gone grey by then?"

"No, I'm not sure why you seem so disappointed to hear this, but he had red hair for quite a long time. Practically until the day he died, people talked about his bright red hair."

Trevor tried to imagine the black-and-white photographs he had seen at the museum as they would look in color, with Clifton's apparently famous red hair. Then he remembered a thread that he and Beth had forgotten.

"What about his *beard*, Maggie?" he asked.

"Well, that was *darker* in color for sure, and it did at least start to grey. The painting of him and Heather shortly after their wedding that we have upstairs shows hints of grey, and the one in the attic even more so. That's good is it?"

"That depends," Beth said with a smile, "on the answer to this last question on the topic. When did he *shave* his beard? I mean, he had it for a long time and then later in life it seemed to be gone. When did he stop wearing it?"

"That's easy. I remember great-grandmother's entry on the first time she saw him without the beard. She wrote about Heather's funeral and she commented about his lack of a beard because she thought he looked awful without it, but was very impressed. She viewed it as a sign of mourning, you know, as a righteous thing he did. Why, what do you think it means?"

"Wilson told us that when the police found Heather, she had some short grey hairs in her hand. It was just about their only clue to who her murderer was ... someone with short grey hair, or a grey *beard*."

"So," Maggie's eyes lit up in a way that made her look years younger, "you really think Graeme Clifton was a murderer! How exciting! It makes sense, I suppose. She must have pulled out a reasonable portion of his beard in the fight for them to notice it in her hand. At the funeral he would have looked suspicious if he'd come with one bald cheek, so he just shaved the entire beard off. That certainly makes sense ... but it is not really proof, is it?"

"I doubt we could ever *really* prove it after so many years, but there is one important piece of evidence left unexplored. Wilson also told us that he found some math papers among Heather's personal effects. I'd like to look at them, if I could." Trevor drank the last drop of his tea and put down the cup, impatient to see the mathematical scribbles that could finally resolve the mystery.

"Why didn't you tell us about those papers, Maggie?" Beth asked, wishing she did not sound so accusatory. "We asked you about any evidence that might show that Heather was a mathematician."

"I'm not hiding anything from you, dear. This is the first I'm hearing of them myself! To be honest, I let Webster Wilson look at her things up in the attic, but I never looked closely at them myself. That is, although I've opened the box and seen the purse and clothing inside, I have always been rather put off by the idea of the murder and have never been able to get myself to examine them further. But, with your help, and a better understanding of what happened to poor Heather all of those years ago, maybe this time will be different."

The young couple was very impressed with how well Maggie did walking up all of those steps to the attic, but the only impressive thing about the attic itself was how unimpressive it appeared. They supposed, though they were no experts on antiques, that they should be impressed by many of the things that were up there. Each of the floors they passed on their way up made them gasp, exchanging glances of amazement at the beauty and grandeur of this old house. Yet, the attic looked remarkably like all of the other attics they had seen in their lives: dusty and dark, unfinished and full of old forgotten bits of the past.

Near a small window that looked out over the street where they had been waiting earlier was a large oil painting of Heather and Graeme Clifton, with a beard that was definitely looking quite grey. *Perhaps*, Beth thought while looking at it, *the beard was really totally grey, and the red tint given to it in the painting was merely done to suit the artist's aesthetics or the subject's vanity.*

Beside the painting, a plain brown box contained the clothes that Heather Blaine Clifton was wearing on the day that she was killed. Her large satin purse, matching the dress and a ribbon on the hat, was caught on a nail head, and it tore as Trevor took it. Inside, as Wilson had promised, they found pages of handwritten notes.

Trevor brought them to the window and looked quickly through the yellowed pages. Even to Beth who had no appreciation of the meaning, there was beauty in the pages of neatly written mathematical symbols. Trevor sat down on an aged ottoman, releasing swirling clouds of dust into the air, and looked through the pages more carefully. Nearly ten minutes passed in silence. Most of the time he was shaking his head and blinking rapidly.

"I don't believe it," he finally said. "This looks like it's really it. I mean, this is the rest of Clifton's last paper ... including the proof. I think it's true! This is the *proof* of Clifton's conjecture!"

"Are you sure you should be calling it that?" asked Beth.

Trevor looked up at the painting of Heather and Graeme and noticed the date painted below the artist's signature. It said "1868."

Only two years later, Heather's body would be lying in a London police station. Somehow, the sparkling eyes of the woman in the painting made this hard to believe. It almost seemed if they must still be alive . Heather's portrait smiled at him as he stood up and held the pages out towards her painted hands and said "You're right, Beth. I should have said 'This is it ... this is the proof of *Heather Blaine's Theorem*'!"

A chill ran down Maggie's spine and she whispered to herself "Amazing."

3

The Adventures of Topology Man (Origin Issue)

It was Cathi who told me to take topology with Professor Smegman. I couldn't refuse. I'll admit, she is quite a nerd. Although her complexion is clear, Cathi's skin is too pale. Her limp blonde hair hangs flat around her face and is cut straight across her forehead in bangs. Her eyes are always red and watery from dust mite allergies. She listens only to classical music and has never ridden a roller-coaster. And, the only thing that really makes her smile is talking about math. But, when she *does* smile, there's a glow that ... that I wouldn't dare attempt to describe. I saw it once when she gave me some advice about my calculus class during freshman year and I could still see its afterimage when I closed my eyes a week later. I knew I had to see it again. So, I talked to her some more about math ... and some more ... and some more. The next thing I knew, I was majoring in mathematics even though most of my friends—and my professors, unfortunately—think I'm more suited for a career as a P.E. teacher.

Then she told me to take topology with Professor Smegman. "You'll like him," she said, "because he really tries to relate to the students."

That sounded good to me. The little bit I'd seen of topology so far seemed a bit obscure, and I was starting to fear that mathematics was getting too abstract for me. So, I took her advice and signed up for Smegman's class as she advised.

I guess it is true that he really does try to relate to the students. On the first day of class, Smegman—an incredibly short man with a perfectly round bald spot on the top of his head—wrote the title of the course on the blackboard in neon colored chalk: "Getting Jiggy with Topology."

We read short stories like "Paul Bunyon and the Conveyor Belt" to learn about Möbius Strips and watched some video about a sphere being turned inside out that didn't make any sense but was kind of cool.

He encouraged us to write haikus about the illustrations in our textbooks.

He referred to his graphing calculator as his bling bling.

When I went to him for help outside of class, he greeted me by saying "Hey, homeboy!" and offered me a free Gatorade from a vending machine that he keeps right there in his office.

He really *tries* to relate to the students, but it did nothing to help me understand or appreciate the abstractions of topology. I still failed the first test and was not too optimistic about my chances on the second.

And then something happened that really changed my life.

I was working part-time in a physics lab with a professor who was attempting to detect particles he called "monopoles." All I had to do was keep the detection machine going for a few hours a night, something I could easily do while getting my topology homework done at the same time. Every once in a while I had to press a button, give it a shot of coolant, and check the monitors to make sure that nothing was going wrong. After a couple of weeks of this I got a bit too comfortable.

Sitting with my feet up on top of the detector, a copy of Munkres' *Topology* in my hands and Guillemin and Pollack's *Differential Topology* behind my head as a sort of pillow, I was startled by the sound of an alarm. First I noticed that the monitor was flashing warning messages like none I had seen before. It announced "tachyon decay" and "monopole detection success", and only then did I notice that my feet were completely numb and that the textbook in my hand was glowing.

Of course, I was embarrassed and quickly hid the books in my backpack before the professor came running in. He looked briefly at the monitor and howled like a wolf. This result made his day, or maybe even his decade. But I sheepishly excused myself and headed straight for home.

I'll tell you, it isn't easy walking with numb extremities. The feeling in my feet slowly returned, but as it did, a tingling sensation spread throughout my entire body. I felt awfully strange, and it probably showed. The odd way I was walking had allowed some street corner thugs to single me out as an easy mark ... or so they thought.

Before I knew what was happening, I was surrounded by a handful of tough kids. One with a knife ushered me into a dark alley where they demanded my wallet. I slowly took off my backpack and *honestly* had every intention of reaching in and pulling out my wallet, but for some reason grabbed Guillemin and Pollack instead.

It fell open to the page I'd been studying, the one I was having trouble grasping. It said:

An orientation of X, a manifold with boundary, is a smooth choice of orientations for all of the tangent spaces $T_x(X)$. X is orientable if it may be given an orientation.

Reading it again in this bizarre circumstance, the point of the definition was clear to me for the first time. Orientation had to do with knowing which way you were facing, like a hiker whose compass always lets him know which way is North. If there were spaces that we call *orientable*, then there must also be ones that are *not* orientable! I began to wonder what it would be like to be inside a space whose very topology prevented you from orienting yourself. If only ...

"Come on, you frickin' nerd!" one of the hoodlums shouted, pointing his outstretched arm and knife straight towards my throat. "Give us your wallet *now!*"

Suddenly confident, I stood up straight to my full height and crossed my arms. One of the muggers became a little nervous now that I did not look quite so helpless, but the one pointing the knife at me was simply infuriated. He lunged straight towards me. By the time he got to where I was, the knife was pointing way off to the side. Unsure of how he could have missed me, he took a step back and lunged at me again.

"Come on and take it from me," I taunted. "If you can figure out how!"

Several of them now tried to crowd in on me, but with each step they took their sense of disorientation grew. Two of them stood still to avoid the uneasy feeling, but one continued to try to crawl towards me. He tried unsuccessfully to grab my pants leg and ended up flat on his back, moaning in frustration.

Somehow, my experience at the lab had given me super-powers ... the power to manipulate the topology of the space around me! At this point, I simply walked away. Leaving the young thieves in the alleyway and returning the normal topology of space as I escaped.

Now, I have read enough comic books to know what I had to do. Once I was home, I got out an old blue sweatsuit, a shiny red windbreaker, a pair of scissors and my sewing kit and I made myself a costume. The red helmet and vest each bore mathematical symbols, matching red holsters on my hips kept the textbooks in easy reach, and the blue material was just tight enough to allow me all the freedom of movement I would need to fight crime and promote good in the world as ... *Topology Man!*

Fortunately for me, my Latin class the next day was cancelled. Professor Wheelock had left a note on the door explaining why he couldn't meet us. It read *Hostes alienigeni me abduxerunt.* I wasn't sure what that meant in English, but I was happy to read it. I wanted to try to find another opportunity to utilize my new powers.

After a few hours, I began to doubt the correctness of the path I had chosen. I was getting strange looks from people as I patrolled the campus, and had not spotted a single problem worthy of my attention. I started to wonder whether being a super-hero was not as easy to do in real life as the comic books had made it seem.

Then I noticed a dangerous looking character. He was lurking in the shadows behind the new parking deck and—most suspiciously—wearing a t-shirt from Hollyhill College. He became tense as a black Jaguar pulled up to the exit.

The driver's window rolled down and the Dean of Undergraduate Studies stuck his arm out of the window. In his hand was his wallet, and in his wallet was an access card which let him in and out of the parking garage. As he held his wallet up to the card reader, the suspicious individual dashed from his hiding place, grabbed the wallet, and ran straight towards me.

"Help," shouted Dean Harrison. "Someone has stolen my wallet!"

At first, I attempted to block the fleeing criminal the ordinary way, like a football player. He easily dodged me and kept on running towards the old library building. So far, I was not worthy of my costume.

So, I reached into my textbook holster and pulled out Munkres. It opened to page 98 where I read:

A topological space X is called a Hausdorff space if for each pair x_1, x_2 of distinct points of X, there exist neighborhoods U_1 and U_2 of x_1 and x_2 respectively that are disjoint.

I remember when Smegman lectured on that, and it made no more sense to me now than it did then. I knew that Hausdorff was interested in ruling out some sort of weird topological spaces, but what did it have to do with my situation? Then, I began to see it. In a Hausdorff space, like the one we live in, the bad guy (x_2) can always avoid the superhero (x_1) because he can get into a disjoint neighborhood out of his reach. But, if this was the right sort of non-Hausdorff space, he wouldn't be *able* to avoid me.

All I had to do was realize it and it came to pass. Try as he might, the thief found he could no longer run around me. Wherever he tried to go, there I was, blocking his way!

He was getting frustrated and tired, and I could do this all day, but I needed a way to finish the job more quickly. I could hear the sound of cars honking their horns, probably stuck behind the dean who was unable to get out of the garage without his wallet.

So, I reached for my other weapon. Guillemin and Pollack fell open to page 84. There I saw a picture of what appeared to be a pair of pants and a description of what it means for two submanifolds to be *cobordant*. I had to think fast here, we had not done the problem in which this definition appeared and so I had never seen it before. What was this example supposed to be showing? The waistline on the pants was labelled X and the cuffs of the legs together were Z ... and they were cobordant because the pants connected them. This was unusual, however, because normally you'd expect the ends to be the same topologically. Ah!

The wallet thief was winding up for one last attempt at getting around—or *through*—me. I had to act fast. I quickly restored the Hausdorff property of space and tinkered with the cobordism of his trousers. In particular, I changed the two cuffs on his pants into just another copy of the waistline. Of course, he tripped, losing his grip on the wallet which landed just a foot in front of me.

I casually picked it up and jogged back to the garage entrance where Dean Harrison waited with an increasingly angry backlog of commuters waiting behind him.

"Thanks," he said as he took the wallet and waved it in front of the reader. "Who *are* you?"

"The name is *Topology Man*, and there is no reason to thank me. It was my pleasure to be able to see justice done today, Dean Harrison."

The next day I felt great. I received an A+ on my translation of Julius Caesar (it was his weird letter to the Senate about the unicorns in France.) I understood Smegman's lecture on Brouwer's Fixed Point Theorem (except for the interpretive dance at the end, which just looked like a drunkard's walk). I had front row tickets to the afternoon basketball game against Hollyhill. And—best of all—Cathi asked me out for dinner!

Well, it wasn't like she asked me out. But, she said there was something she wanted to talk about and I was the only one she could trust. It looked like it might be something unpleasantly serious, but I was really looking forward to it.

How could I have known how bad things were to become?

The basketball game began, as tradition would have it, with the two teams together pushing The Trophy into the center of the court while the bands each played their school's fight song. The Trophy, a four foot tall golden cup, was just a symbolic prize for the winner, passed back and forth each year according to the outcome of the game. Our team had not won it in a few years, but we were all confident that the game today would change that.

Even before The Trophy was pushed off-sides, a short man in a black leather suit and green mask rushed onto the court. He had "super-villain" written all over him. He pointed at the wooden floor and it deformed so that a ten-foot wall of wooden planks imprisoned the basketball players, leaving The Trophy unprotected.

Fortunately, I had worn my new uniform under my clothes. I was able to get to the restroom, put on my mask, take off my street clothes, and return before the bad-guy was gone.

He was now in the process of shrinking The Trophy down to a tiny size. Once it was the size of a sugar bowl, he grabbed it and headed for the exit. I jumped down onto the court in front of him and yelled "Where do you think you're going with that Trophy, leather-face?"

"A new super-hero here to battle the power of Homotopy ... isn't that cute?" he shouted back. "Too bad you won't be able to do any-thing when you're imprisoned inside a giant, everted basketball!"

At his command, the basketball near my feet began to grow. Once it was taller than me, it twisted around like the sphere in the movie and it ended up inverted ... with me trapped inside!

It was pitch black inside the basketball, and the sounds of the crowds outside were muffled. I wasn't sure how much air I had in there, but it was certain that my time was seriously limited.

Fortunately, I had thought to include a pen-sized flashlight in my costume and was able to use it to browse page 366 in Munkres. There I saw the Klein Bottle, a topological object that had much in common with the sphere which imprisoned me: it was a compact surface without boundary. There was one big difference, however. Because of its unusual topology, the Klein Bottle has no inside or outside. Every point in space is on the same side of the Klein Bottle.

Endowing the basketball with the topology of a Klein Bottle, I was able to escape in time to see Homotopy running out of one of the exits. Just before he vanished from view, he turned back to witness my escape and stopped.

"A topologist, eh?" he said, sounding pleased. "That makes you a very worthy opponent, indeed. But what can you do about *this*?!?"

As he said "this," the floor of the basketball court began to deform again. It returned to being flat, but now each point was continuously shifting and moving around, making it almost impossible to stand. This problem was too easy to solve, however. I didn't even need my books. It was just earlier that day that I had seen Brouwer's Fixed Point Theorem which guarantees the existence of a place on the floor which was not moving. I leapt into the air, did a back flip and landed smugly on the unique fixed point.

"Brouwer's fixed point, huh?" whined Homotopy. "You rookies are so *predictable!*"

It was only then that I noticed that this fixed point placed me directly beneath one of the hoops. Before I had a chance to move, Homotopy shrunk the screws which held the hoop to the backboard, enlarged the hoop as it fell, and then returned it to its normal size once it was past my shoulders and chest.

With the hoop holding my arms tightly, I couldn't reach my textbooks. Moreover, I could barely breathe. It looked as if I wasn't going to be able to make it to dinner with Cathi after all.

But, just as I was about to give up hope, a woman in a sky-blue leotard fell from the banner hanging over the exit and flattened Homotopy.

He looked up at her and said "Category Theory Girl! So, we meet again. Only, this time I am the functor."

"Only a functor of *evil*, Homotopy!" she shouted. As she whipped out a short exact sequence and attempted to map the villain into the kernel, I couldn't help but notice the fringes of straight blonde hair sticking out from below her mask. I thought that I was saved when she flashed that amazing smile and said "I've got you figured out, Homotopy. Your morphisms are nothing but continuous functions from I to $M \times I$!"

"That's what *you* think, homegirl!" The floor beneath him morphed into the shape of a tall throne, with Homotopy sitting high above her. He clapped his hands, and the thunderous shock wave that resulted knocked Category Theory Girl to the ground.

"Uhnnn," my erstwhile savior moaned. "A 2-functor from Top to Vect ... but *how*?!?" Then she collapsed into unconsciousness.

Could this spell the end for Topology Man and Category Theory Girl?

Be sure to read the continuation in the next exciting issue of Topology Man!

4

Eye of the Beholder

I don't know how long I've been like this. It could be a week, a year, or just a few minutes. Time means nothing to me now. Perhaps it was just a few days ago that I went through my usual daily routine for the last time.

The TV turned itself on automatically at 6:30 AM, signalling to me that I had exactly a half hour before I really had to get out of bed. The newscasters told me about the traffic and weather. I half listened, knowing that in between telling me why I should watch this channel again at 5 PM and announcing the wonderfully famous people who would be guests on the late show that night, they might possibly inform me of any Earth-shattering events that had occurred since I last saw them at 11:30 PM. It didn't matter. I didn't care about any of it anyway.

By 7 o'clock, I was in the shower, looking forward to the only things that could get my jaded psyche feeling much of anything at all: the numbers and their mysteries. Without even bothering to dry my hair, I just threw on enough clothes to keep me from getting arrested for indecent exposure, grabbed a chocolate Pop-Tart and a drink, and headed out the door.

I set the Pop-Tart on the dashboard of my tiny green hybrid car to be warmed by the bright morning Maryland sun, my Coke in the cup holder, and got just a little bit of pleasure from the squeaking sound of the cheap plastic seats as I adjusted myself for the long commute into work. Aside from that squeak, the car was practically silent.

When I arrived, the guard at the parking lot tried to start a polite conversation with me. He was the only one at work who bothered anymore.

"Good morning, honey," he said. I didn't really mind him calling me 'honey'. It wasn't a come-on or anything. He seemed to think he was my surrogate father or something. "Didn't I tell you to start eating a better breakfast?" he'd continue.

He must not have been privy to the terrible things my co-workers called me. Most people, I suppose, just thought of me as weird or pathetic. They had become accustomed to my apathy and lack of "social graces" and attributed these oddities, strangely enough, to my genius. Others, perhaps those wise enough to realize that I was no genius, just called me "the bitch." It's funny, because nobody who knew me well before I came to work there would ever have called me either.

The truth is, I was simply depressed. Terribly and utterly depressed. I was so depressed that I didn't even want to figure out how not to be depressed anymore. The thought of not being depressed was as depressing as anything else. I guess I'm over it now. Who would have thought that when I finally got over my depression, I'd remember it longingly?

As always, the first thing I found on my desk was a packet of encrypted intercepts that I had to work on. A little bit of effort from me, and these strings of unintelligible digits would turn into messages, communications from one person to another. On rare occasions, the messages were of obvious importance, even to me. Sometimes they contained too many code words for me to grasp their meaning. And, sometimes, they were clearly nothing secret at all, just an ordinary message, a shopping list or love letter, that happened to have been sent encoded and deemed "suspicious" by someone in the government.

This was not what I lived for, but it was still a good warm-up for the day. I suppose, if I wanted to be really psychoanalytical about it,

I could say that turning the apparently random number strings into meaningful messages gave me hope that there was some meaning to it all, hidden behind the apparent cruel pointlessness of my life. Perhaps to a small extent that is even true, but then, as now, I am almost certain that no such meaning exists, and am absolutely certain that even if it did, I wouldn't want to know about it.

Really, the only thing I liked about these morning exercises was that they were little puzzles, giving me something to do with my brain and leaving me with a sense of accomplishment when I was done. Each one required at least a bit of clever thinking to decode, a special nuance here or there. I never knew what they were going to throw at me next, and which little piece of information, whether it came from number theory or some non-mathematical discipline, I was going to have to dig up. But, there was nothing terribly challenging involved since the hard work had already been done for me. In fact, the people who sent the messages probably did not worry too much about the extra twists and gimmicks that they added to protect their secrets since they assumed that we would not even get that far without the key. And it was only the twists and gimmicks that I was working on, since the key was given to me. Each of the messages came with the key attached to it, handwritten on a little yellow slip.

The keys to turning these number strings into messages were numbers as well, very special numbers in fact. To decode them, one needed to know the really huge number that—along with a little bit of modular arithmetic—was used by the author of the message to encode it. The thing that made these numbers special was their lack of factors. If you make up a 20 digit integer, chances are that you could divide it evenly by some smaller number. If the last digit happened to be a 0, 2, 4, 6, or 8, for instance, you could divide it by 2. Or, if you take the individual digits and add them together, repeating the process until you end up with a single digit, and that digit is a 3, 6, or 9 then the number is divisible by three. It would be a rare number that is not divisible by any of the integers less than it, but these keys are precisely such numbers. To put it bluntly, they're prime.

The thing that I was *supposed* to be spending most of my time on in this job is figuring out a way to look at the coded message and from it deduce the key. Once I was done with the intercepts for the day, I was free to spend the rest of my time on that project, with only occasional interruptions for paperwork and other useless stuff. Of course, many brilliant mathematicians over the years have explored the properties of prime numbers, and recently they have been the focus of intense study because of their role in cryptography, and nothing in the whole of mathematical knowledge gave me any reason to think that such a task would be possible. Nevertheless, my job was to do it.

At first, I was working in a complete vacuum. There was no guidance about how or where I should begin my investigation. During this time, I did my best, but found it difficult to put much effort into it, both because of my depression and because of a belief that the task that had been set for me was not even possible. I suppose now that it was precisely to get me working harder that I soon started receiving strange "hints" from my bosses. The frustrating thing was that these hints seemed to suggest that they already *knew* the answer! For instance, they began giving me not only intercepts that I was expected to decode, but also others that did not come with keys. These, I was told, were different in some way that would prevent the algorithm I was supposed to discover from finding the keys, and I was only supposed to determine in what way they were different from the others.

This was the mystery that *really* kept me living from day to day. Why, if they already knew the answer, were they so interested in having me discover it? I sometimes even felt as if the whole thing was an act that they were going through for my benefit. As if the President himself had ordered America's top security agency to create a puzzle that would entertain a poor, depressed mathematician and keep her busy for just a while. Of course, I knew the people working above me too well to think that they cared much about any of the staff working for them, but it was a nice fantasy.

The obvious *real* explanation, I thought, was that our government had found a device that was able to do it! They had bought, stolen or otherwise obtained a device that could actually find the keys hidden in the coded messages themselves. The fact that I knew the answer was *out* there motivated me to work harder, which I guess was the point. But it also gave me another mystery to solve: if they had this thing, why did they need me for anything? Did they not know how it worked? (It's hard to imagine such a device of *Earthly* origin at least that they couldn't just analyze to determine the algorithm without my help.) Did they know how it worked, but want me to discover it so that they could *claim* to have figured it out independently? (This was my least favorite scenario ... I would have hated to be involved in such a scam.)

Further hints soon made it clear to me that they had a device that they couldn't understand or duplicate and that they did not expect to last much longer. At first, superiors casually mentioned 'Carl' to each other in my presence. Since it was clear that 'Carl' was determining the keys for them, I supposed that this was an acronym for something ... Cryptographic Analysis R-something L-something. Later, they would talk directly to me about it, as if I knew what they meant.

"We've got to hurry," they'd say to me. "Carl is not going to be with us much longer."

"We cannot stress the urgency of this situation enough. Our national security has grown to depend on Carl and we would hate to think of what would happen if you do not have a way to emulate the algorithm soon."

In keeping with my sullen genius persona, I would usually just nod or grunt to acknowledge that I heard them and then get back to work. But, one day, not too long before it all ended, I lost my cool and exploded. "Why don't you just let me look at this Carl thing before it dies on you?!? Perhaps if you let me actually *play* with it myself I can figure it out easily."

"It's not that we haven't thought of it, Bev," they admitted to me, "but we were saving it as a sort of last resort. Still, perhaps it is time. We'll have to check and get back to you."

A few days later, they did. Jim and Sharon called me to the conference room. I'd only ever been there before with a huge crowd of

people, and it was somehow frightening to be up there with just the
two of them. He sat at the head of the darkly stained mahogany
table and she was next to him. I wasn't sure which of the heavy
cushioned chairs to take for myself, but finally worked up the nerve
to take the seat right across from her. When I did, she picked up a
remote control on the table and used it to deadbolt the door to the
room. Putting the remote down again, she glanced momentarily at
another door at the back of the room, one I don't think I had ever
noticed before. Or maybe I had noticed it and ignored it, assuming
it was just a broom closet.

"You do not technically have clearance to see what you're about
to see," he said, pulling out a red folder. "And we don't have the
power to grant you such a high level of clearance."

"However," she said, picking up where he left off, "we have gotten special permission to let you meet Carl now."

"We both have studied your history and psychological profile
carefully, and believe that you are now at a stage where we can trust
you with this."

He said no more, but waited for me to nod my agreement before
handing me the red folder. I had time to read just the label on the
front that said "Rockford, Carl Walter" before she used the remote
to open the door at the back of the room and I caught a whiff of
that unforgettable smell.

M y parents tell me that when I was a kid, it was clear that
I loved three things. I loved drawing pictures, and
though I was no Picasso, I always seemed to be drawing
pictures that were more advanced than the scribbles my classmates
and older siblings could achieve. I loved my stuffed animals, which
I always matched up into families ... mothers, fathers and babies.
And I loved my math homework. I not only did the math homework first, because I couldn't wait to start on it, but I always pulled
it out again at the end to show to my parents, their friends, or my
stuffed animals so that I could explain to anyone who would listen
"how it worked."

By the time I was in high school, I no longer had any stuffed animals, but I still loved art and math classes. I had decided, in fact, that I was going to be an artist and was very pleased when I got into a prestigious college that was known for its graphic arts program. Unfortunately, it didn't live up to my expectations. Even though I liked what I was producing, all of my art professors seemed to hate my artwork. They said that the skill was there, meaning that they appreciated my draftsmanship, but that I needed to put some beauty into it. Isn't beauty in the eye of the beholder, though? I thought my pictures were beautiful. It seems that I just had the wrong sense of aesthetics when it came to art, and I started to doubt that I would be able to get through school if this continued.

For the first time in my life I felt completely lost and directionless. The first time, but not the last.

All the while, I had been taking math classes, both for fun and because even art majors have to fulfill their math requirements. Of course, I was the only art major who was taking the honors advanced calculus class, but it still had not occurred to me that I could *major* in math. Art was something I thought of as a possible career—everyone knows that there are people who are artists—but I had never been introduced to the idea that math was anything other than a class you take at school, albeit one that I enjoyed.

Then, I somehow ended up attending a meeting of the Undergraduate Math Club at which a speaker came from a neighboring university to talk about his research. His research was probably really interesting, though I'm not sure I understood most of what he said, but I did understand a few of his remarks about math in general. What I really learned from his talk was that you could get a degree in math, and get a job at a university where you actually do your own math research or a job in industry where people who don't like math will pay you a lot of money to do their math for them. I no longer was lost; that same day I filled out the declaration form and became a math major.

It wasn't until I was already working on my PhD thesis, four years later, when the unfortunate irony hit me. I had run to math because I found myself being persecuted for my "bad" aesthetics in the arts. This seemed to make sense to me for a long time, since math was an objective discipline in which you were either right or wrong ... aesthetics had nothing to do with it. Or so I thought.

In fact, mathematics is as aesthetically driven as any human endeavor. Other mathematicians might agree on when you've proved a theorem, but unless that theorem is perceived as being beautiful in some way, you won't get a lot of appreciation for it. And, with fate conspiring against me as always, it should be no surprise that I have the *wrong* aesthetic for mathematics, too. Just like my art, described as technically correct but aesthetically lacking, my math research was not really appreciated by my thesis committee.

One of the main sources of beauty in mathematics seems to be derived from balance between the difficulty of the problem and the simplicity of its statement. Most mathematicians, it seems, find math especially beautiful when a very simple statement turns out to be unbelievably difficult to prove or disprove. Fermat's Last Theorem and Goldbach's Conjecture are two examples of mathematical statements that are beautiful by this criterion. Any elementary school kid can understand the questions, but not even Gauss or Hilbert could answer them.

My tastes, unfortunately, run just the opposite way. What I love in math is finding a terribly difficult sounding situation, something that you would initially guess is simply beyond our comprehension, and finding out that it is actually much simpler than anyone else ever imagined. Here's an example: my thesis advisor had developed a new class of vertex operator algebras for addressing a non-commutative analog of Goldbach. We were supposed to be studying and applying them together. It took me a year just to grasp the definition of the damn things; my advisor and I were the only two people in the world who understood all of the conditions that defined these new algebras. However, before we got very far on the applications, I proved that these apparently complicated structures were actually isomorphic to something elementary that all first-year grad students see in their algebra classes. All of the complicated conditions could be reduced to a few elementary ones in a non-obvious way. To me, of course, this was beautiful! I had found that this very messy thing, when viewed just right, cleaned itself up into simple order. To her, however, it was a disaster; it meant that the whole idea had to be scrapped as being too trivial, and that I would have to find a new thesis topic.

Once again, I felt completely hopeless and directionless. My advisor wanted me to find a new topic, even though I would have preferred to simply publish what I already had and be done with it. Find a new topic?

I found a new topic, all right. His name was Brad. We met at a birthday party for a mutual friend and hit it off right away. He was gorgeous, and fun, and smart, with an English accent and seemed to think all of the things I hate about myself were "adorable." Within a month we were living together, though I still paid rent on my empty little apartment, and were oh-so-much in love.

Brad had already earned his PhD in theoretical biology at Cambridge and was here on a postdoc. Boy, did he work hard! I mean, when he wasn't working he was a joy to be around, but when he *was* working he was *working*. I could never be like that. Without his support, I'm sure I would never have gotten a PhD in math at all, but he helped me deal with my advisor's bullshit. Though I never really thought it was better than what I'd already done for her, I did what she asked. She wasn't that happy with my attitude either, but she had to admit that I did what she'd assigned me and so, at long last, I got my degree.

To celebrate, we went away for the weekend to a B&B in the country. A beautiful old house, with creaky wooden floors, scroll-work on the porch, quaint embroidered phrases framed on the walls, and the most amazing huge shower in our private bathroom. We really had a great time in there! Well, I did anyway. One time, as soon as he got in with me, I aimed the handheld shower head right at him and nibbled on his neck. "Oh, that's hot!" he said. "Oh, yeah, hot" I said, misunderstanding. He jumped out of the shower, falling ungracefully on the mat outside, knocking me over with his shoulder, and giving me a bit of a black eye in the process. "No," he said, "I meant that it was *hot*!"

Neither of us was really hurt, and it was sort of funny. Brad mentioned that he had read that extreme heat had some contraceptive properties, and we decided to test it out, by making love without the use of condoms for the first time.

You can't tell much from a single trial of a scientific experiment, but we did learn one thing from this experiment. We were both fertile ... perhaps too fertile. By the following year our lives were completely different. We were married with a very needy little baby boy and less than impressive jobs as assistant professors at a small liberal arts college in Virginia. It was not exactly what I had imagined for myself, but we were really happy. I didn't have that lost feeling, anyway.

C arl's place had the unmistakable aroma of a sick room. Growing up in a home with two kids and two invalid grandparents, I knew that smell well. Of course, none of the sick rooms I'd ever been to were accessed only through a secret door at the back of a conference room at the headquarters of the National Security Agency, and none of them had been guarded by a marine with such a scary looking rifle.

Carl, sitting up in his bed staring at a blank wall, appeared to be a young teenager, with the first signs of facial hair making their appearance on his pimply chin. A glance at the information on the first page of the red folder confirmed that he would be turning 16 in a couple of months. It also said that he was from Cedar Rapids, Iowa, and so far did not give me any explanation of why he was in Washington, DC.

I was a bit surprised by the fact that there were no greetings exchanged, and that nobody introduced me to Carl. When I was just about to ask about this, Carl shouted "Two thousand three hundred and eighty nine! Six thousand five hundred and twenty nine!" He giggled, and then there was silence apart from the humming of machines in the room. I was not, and still am not, sure whether this had anything to do with our appearance or whether Carl would have shouted this out alone in his room, in any case. Even if he had, it would not have gone unnoticed, since I'm sure that the many cameras and microphones clearly visible around the room would have recorded it for later analysis.

I was offered a seat and witnessed what I suppose must have been the precursor to all of my work here at the NSA. Carl, who previously appeared to pay no attention to us at all, seemed very interested in the stack of intercepts that the bosses had brought in with them. He looked through them all, one by one. Nearly all of them produced a lopsided smile, and after a minute or so, he would recite a number. The bosses copied them down on yellow slips and attached them to the messages. I had no way of knowing whether the numbers he recited were primes or whether they were the keys to the corresponding intercepts, but I know that in all of my time working here, I always found that the number on the yellow slip I was given was in fact the key to the message it accompanied.

Every once in a while, Carl would frown at a message, and simply pass it by. It did not take him much time at all to make this decision. He'd hardly have looked at the message when he would hand it to the bosses without saying a word. I guess he just didn't like those for some reason.

While this went on, I flipped through the contents of the red folder and learned a bit about Carl's history. As a child, Carl was considered to be unusually intelligent, nothing that captured national attention or anything, but his parents were very proud of him, and he showed a lot of interest in mathematics from the time he was five. Then, when he was eight years old, his parents found him in his room in what they imagined was a coma. They rushed him to the local hospital, where the doctors found him to be physically healthy and were confused by his behavior. When he was transferred to the university hospital in Iowa City, doctors diagnosed him as having a very severe form of autism. This did attract national attention, because nobody had ever seen autism with such a late onset. He was examined by many doctors and psychologists seeking to understand autism. For the most part, they were doctors who had no appreciation for his habit of reciting numbers, but nevertheless they noted and occasionally made some mention of this phenomenon. Finally he was "interviewed" by two psychologists employed by the NSA ... my bosses. Shortly thereafter, he was diagnosed with a brain tumor and, according to the information in the folder, quietly passed away in the hospital at the age of eleven.

B rad and I both found it difficult to do our research at our new jobs. Between our heavy teaching loads, our academic isolation, the poor library facilities at the college (at least when it came to science research) and taking care of the baby, we found ourselves short of all of the resources we needed. So, during one of our dinner time discussions that so often found us talking to each other about math and biology, we came up with the idea of pooling our resources and working *together*.

If we'd been at a major research university somewhere, we would never have done that. I'd have been working with the other number theorists in my department on my quantum numbers and Brad would have been working on his "top down architecture" theory of ecosystems, mine completely free of any biological applications and his completely free of any non-trivial mathematics. And, that would have been a shame, because as it turns out, our work together was a lot of fun.

It all started as a joke, when Liam first started going to daycare, about how we had to worry about all of the new germs he would be encountering. "Germs cause your mucus membranes to overproduce, cause your digestive tract to reverse its direction, leave you feeling sick in bed. Why", I asked without the least bit of seriousness, "aren't there germs that do *good* things?"

Neither of us gave this question much thought until the next time it came up again as a joke later in the week. But this time, it dawned on me that helping the person that it infected could be beneficial to a bacteria. I mean, why should it *want* to make the person sick? If they stay healthy and live longer, that could give them more of a chance to reproduce and spread and that's all they're about.

It took Brad almost a whole minute to realize that there *are* such bacteria and science already knows about them: commensal bacteria are the bacteria that live in our body all of the time. Not only do they not cause disease, but they are helpful or even necessary in normal circumstances.

However, Brad was not able to think of any *viruses* that were helpful to their human hosts. And that, it turns out, was the start of our collaboration. By the next afternoon we had done some research and found the relevant information that appeared in the few books and journals we had access to at the College. By the next month we had gotten even more information through inter-library loan and internet sources. And within a few weeks of that we actually had something worth showing people.

Our paper began with a mathematical analysis of the situation from an evolutionary standpoint. Using the standard formulas for the development and spread of an infection used by virologists, and adding a few extra tweaks here and there, we were able to show that there is a threshold beyond which aiding their host is actually advantageous to the virus (although the costs outweigh the benefit from the virus's evolutionary viewpoint when only a small amount of aid is considered). As has been shown many times in the history of mathematical biology, if something has been found to be mathematically possible, then there is an awfully good chance that it can be found in reality. So, our next two sections address the question of where such viruses could be found. In a relatively math free section, Brad explains why such viruses might not have been isolated or identified yet by molecular biologists. Then, and this was the particularly gutsy part that miraculously paid off, we analyzed the data on intelligence that had been so controversial for the last few centuries and argued that we could explain it. In particular, although some academics (and rednecks and Nazis, of course) tried to argue that intelligence was determined by race, they were never able to get the data to support these claims. The common liberal response, however, that intelligence tests primarily measured socio-economic status didn't quite fit the data either. Once we considered it as an infection, however, it all started to make sense.

The paper was accepted in a really good journal, although we had to tone down some of our claims of priority when one of the reviewers pointed out that a beneficial virus that gives immunity to AIDS had already been identified. Still, we were thrilled merely to have the paper published. (That was a step towards tenure for both of us.) And we were more than happy with the amount of attention our paper received both in the popular media and in scientific circles.

Almost a year later, when Liam was in kindergarten, we received advance notice about an article to appear in *Nature* that claimed to have found the "intelligence virus" that we wrote about in our paper! By this point, we were already working on our next project—prion cycles—and had practically forgotten about the virus paper.

We were again surprised by the amount of attention this received in the media ... especially since the data seemed rather preliminary to us. But, this probably was an important factor leading to job offers—ones we couldn't refuse—from Georgetown for both of us to work in their mathematical biology unit. And so, with everything seemingly going our way, we moved our little family to a suburb of DC and started a new, exciting and brief life as research hotshots.

The start of the end came when we attended a party at the home of a former Surgeon General. I arrived late with Liam because we had been at a birthday party of one of his friends. By this point he was just starting second grade and had some homework to do. I was looking for a quiet place where he could get started when I noticed Brad, who had already been mingling with the crowd, looking a bit shell-shocked near the dessert table.

"Did you know they've gotten to the point of *giving* the intelligence virus to kids?" he asked me.

"No!" I said. "What, all kids?"

"It was only a limited trial, an FDA approved experiment based on the preliminary data, but they found an unexpected side effect. If the virus is introduced too late, it seems to cause some sort of autism. They aren't sure, but they're worried enough to be canceling the rest of the experiment until they can figure it out."

It was also at this party that I first met the two psychologists from the NSA who would later be my bosses. They didn't know who I was, and hadn't heard about the virus stuff, but gave their usual spiel about needing help from mathematicians. I guess I know now that they'd already been working with Carl for a few years, so I know exactly what sort of help they needed, but at the time I couldn't have cared less about the NSA or those two creepy shrinks.

When it was time to leave, Liam went in the van with Brad. I drove the car that Liam and I arrived in. He must have buckled Liam into his car seat rather quickly, because I already saw them in my rear view mirror as I was making the first turn, and they stayed right behind me all the way to the bridge over the Potomac. At least, I hope he buckled him in. It wouldn't have made any difference anyway, I suppose.

I was feeling pretty good, maybe one too many glasses of Shiraz at the party, and jamming to the radio. I glanced in the rear view mirror just as I was nearing the crest of the bridge and saw Brad's van behind me swerving just a bit. Still looking in the mirror, I saw a big 16-wheeler going the other way behind me jack-knife, fall on its side and slide into the car beside it. That car, that I later learned was driven by someone from the Ecuadorian Consulate, was thrown over the rail by the impact and into the river.

The fact that I didn't hear anything and that I saw it only in the mirror made it easy for me to believe for a moment that it hadn't really happened. Before it consciously hit me, before I realized that I no longer saw the van following me, I was already crying. I actually drove all the way to the other side of the bridge and turned around, but I couldn't get anywhere near the accident. The traffic wasn't moving at all. Leaving my car where it was, I started running between the stopped cars, then just stumbling by and wheezing when I ran out of breath.

The police were already there when I finally arrived. Even though I couldn't find any words, the officer knew from the way I was staring at the crushed van that it, and the mutilated bodies inside, were important to me. He didn't try to guess what happened, but calmly explained that there were no survivors. The driver of the truck, the person in the car that went over the edge, Liam and Brad all died at the scene. The official report that I received a few days later blamed Brad for crossing over the median in front of the truck, but had no explanation for why he did.

For a few days I literally did nothing but cry. That was probably the sanest reaction I had. After that, in a semi-sane state of virtual numbness, and coached by my siblings who wanted me to "get on with my life," I tried returning to work. That, however, proved to be almost impossible. Our house, our offices, even college students who had nothing in common with my dead son other than youth, all of these led me to thoughts of Brad and Liam, and once there I was useless. And so, when the NSA folks showed up in my office one day to offer me a whole different life, I accepted immediately.

"Okay," she said as they left me alone with Carl, "he's all yours!"

This was not at all what I had been expecting, a machine that I could experiment with, a computer program whose code I could try to understand. This was a young boy, even if he was a very unusual one.

I began by simply talking to him. "Hi! Carl? How're you doing?" Of course, nothing happened, but I felt that I had to try. In fact, from the way he completely ignored me I would have thought he was deaf, dumb and blind. However, I had seen the way he reacted to the numbers that Sharon had given him. Not only the coded messages, but also a "treat" at the end for a job well done. He knew it was coming, after the last message. You could see the anticipation as Carl cautiously took the slip of paper from her hand. He read it, considered it, and lit up like a prophet experiencing an epiphany. He cooed and shook, and eventually collapsed in ecstasy at whatever they had shown him. He was still lying there like that while I tried to get his attention. Even when I recited numbers, however, there was no reaction. I suppose that the numbers weren't especially nice ones, I was just making them up off the top of my head, but I thought he'd at least grunt to acknowledge them, or laugh at me for my pitiful numerical taste. But, nothing. I looked down at the folder and noted again that Carl was supposed to be dead. It was no surprise, then, that he wasn't reacting ... he died years ago.

Of course, he wasn't really dead. He was breathing, and moments ago he was enjoying the beauty of some particularly wonderful number. It dawned on me that his parents must not know that he's alive. The death, obviously, was a cover ... a cover for the *kidnapping* that brought him here. My perspective on the whole situation changed with that realization. Perhaps, in my mind, it had some irrational connection to my own loss. In any case, it became clear to me at that moment that this child and his family were more important to me than my job, mathematics or cryptography.

I stood up to leave, looking calm I hoped, and walked towards the door.

"Giving up already?" asked a disembodied voice.

"No. No, not at all. But, I really was not prepared for this. Now that I know what I'm dealing with, I'd like to come back better prepared. Bring some sample crypts, have a few 'pretty' numbers handy, maybe with ..." The door opened automatically and the guard moved aside to let me out.

"Ma'am," he called to me. "I'm afraid that you have to leave that folder in here."

I walked over to the live oak tree next to the duck pond where I usually go to smoke. I probably was walking a bit more quickly than usual, but I made an effort to look casual and did not notice anyone paying attention to me. Just to be certain, I lit up a cigarette and started smoking it, but after drawing on it a few times I got out my cell phone. Of course, they might have moved since, but I decided to call the number that was listed for Carl's parents on his hospital admissions form. There was a chance that they were still there and I just wanted to let them know.

After a few rings, a breathless, Midwestern voice said "Hello?"

"Hello, Mrs. Rockford?"

"Yes, I'm ..."

At that point, the connection was broken. I looked up to see my bosses standing next to my bench, looking very disappointed.

"Bev, you know that you can't do that!" he said.

"And after we demonstrated how much we trusted you. How could you?" she asked. I leapt up and started to run towards the parking lot, but I was tackled by a security guard. Either the fall or some sort of tranquilizer knocked me out, and the next thing I remember is waking up back in Carl's room.

While recovering on the sofa, in a half-conscious state, I had paranoid fantasies about the bosses. There seemed, at the time, to be a sort of logic to my fears. "It is true that I first met them just before Brad's car accident," I thought to myself. "Also, I would never have come to work here if my family had not been killed. Killed? Could they have been *killed* just to set me up for this? Maybe their plan was to make Carl the only outlet for my maternal instincts so that I ..."

It didn't take me long to reject this whole idea as being too elaborate. But, when I finally opened my eyes and saw them sitting there across from me, the hatred I felt for them was more than could be explained by their treatment of Carl. I now blamed them for *everything*. I felt my face flush and my teeth clench, but they sat there calmly.

"Oh, Bev, how could you?" she said again, in the most patronizing way possible.

"Don't you realize what is at stake here?" he added seriously.

"How could *I*? What's at *stake*? You guys *kidnapped* someone's kid!"

"Well, that's sort of a strange way to look at it," he said, wrinkling his brow.

"What's strange about it? You made up some story about him having a brain tumor, told his parents he died, and kept him locked inside a room for all of these years ... sounds like kidnapping to me."

"*Made up* a story about a brain tumor? What do you think we are, Bev? Evil?"

She sounded sincere, and suddenly I wasn't sure I knew what I thought.

"We did not make up the brain cancer. He really has brain cancer, and probably would have died from it years ago if we hadn't stepped in. You would not believe the expense and effort that Uncle Sam has gone to in keeping this kid alive. There's no way this would have happened in some hospital in Iowa."

"But, you took him from his parents!"

"If we didn't, then death would have. There was no way it could have ended happily for his parents. At least this way they had closure and we had a valuable resource. An unbelievably valuable resource that's probably prevented thousands of other parents from losing *their* kids as a result of anti-American terrorist activity. See?"

I didn't see anything clearly just then, including why I had been so angry at them a moment earlier. What they did may not have been laudable, but it no longer seemed to be obviously evil either. So, I just sat silently.

"Look, we trusted you. You said that maybe if you had a chance to talk to Carl you could figure out how he does it. He really doesn't have long to live. If you still think you might be able to do something with him, please do it *now*, okay?"

I agreed to give it another shot. It took a little while until I was strong enough to stand again. During that time, they kept reminding me how important this was to our country. As they spoke this nationalistic drivel, their eyes shone and I could almost imagine that I heard the "Star Spangled Banner" playing in the background.

"Yeah. Okay. I'm going to go talk to Carl now," I said, briefly regaining my famous depressive attitude. "I am really curious to know how he does it." She winked at me and smiled, he patted me lightly on the back, and they left me alone with Carl Rockford once again.

Carl was in the bed on his side. I went around to the back of the bed so that I could see his face, to try to judge his reactions to the things I might say to him.

"Hi, Carl," I began, still uncomfortable with the fact that such pleasantries were not really necessary in this situation. I threw out some numbers randomly, some big and some small, some even and some odd, some I knew were prime and others I had no guess about, but I saw no reaction. Then I tried some equations. Would he like Euler's beautiful formula

$$e^{\pi i} + 1 = 0?$$

Again, no reaction.

I noticed that the hair over his ear was standing up in a cow lick. In fact, it is not at all surprising that Carl had the worst case of "bed head" I'd ever seen. In response to an uncontrollable instinct, I reached out to smooth down his sweaty hair with the tips of my fingers. As I did, he moaned softly.

At least it was *some* sort of reaction, so I kept doing it. He moaned louder, and then he said "That hurts!"

"Carl!? What did you say?"

"It hurts. My neck hurts real bad."

"Carl," I asked softly and incredulously, "can you hear me?"

"Mm-hmm," he said, opening his eyes and staring at me. He sat up in bed, wincing visibly in pain, and asked "Who are you?"

I must have looked stupid, standing dumbfounded by such a simple question, but I honestly did not know what to say.

Looking around the room a bit, he asked "Is this the hospital?"

"Yes," I said, grateful for a cue, "I'm your nurse and this is the hospital. How are you feeling, Carl?"

"I'm not feelin' too good. My neck hurts and my head is funny, too."

"Do you remember what happened to you?"

He sat quietly for a moment, staring at his left hand as he slowly closed his fingers into a loose fist and then opened them again. Then he laughed, and smiled, just a little bit.

"I was thinking," he said. "I was just in my room thinking about numbers. I like numbers; they're cool."

"That's the last thing you remember," I prompted, "thinking about numbers? Specifically, what were you thinking about them?"

"I was just thinking that I bet numbers are how your brain works. You can see pictures and hear words, but I bet you have to turn them into numbers to think about them." He brought his hand up to the back of his neck and began giving himself a massage. "It really hurts; can you get me some medicine?"

"The doctor will be here in a minute," I lied. "He wants me to ask you these questions so that he can treat you properly. Can you tell me more about how you ended up here?"

"Where's my mom?"

I bit my tongue, but smiled, as I told him that she too would be there soon. My eyes began to tear, but Carl did not seem to notice.

He was comforted by the thought that his mother would be joining us there soon.

"Okay," he continued. "I started thinking about how I was thinking. Do you know about *recursion*?"

"Yes, I remember math from school. Recursion is when some definition loops back on itself."

"Uh huh. I liked it because it was like recursion ... thinking about thinking. I was trying to think of how the rules of thinking could just be rules of numbers ... and then I *saw* it."

"Saw what?"

"Did you ever look really close at a comic book? The pictures are just lots of little dots. They don't look like pictures of superheroes at all. They're really just dots. The same thing happened to my room, but instead of dots, it was numbers. For a little while it was cool seeing nothing but numbers, but then I wanted to move or call my mom or something, and all I saw was numbers still. It was like I was drowning in numbers. Some of the numbers were pretty and some were ugly, but there wasn't anything else. I'm glad that ... Ow! ... I'm glad that I'm not just seeing numbers anymore."

He put his head back down on the pillow, and I helped him pull the blanket up to his chin.

"Cold now," he said. Then he looked very deeply into my eyes. I expected him to say something profound, but he just asked "Did I miss a day of school?"

"Yes," I told him, "you've been very sick and have missed a lot of school."

He really grinned when he heard that, and closed his eyes. Then, suddenly, his face twisted in pain and ... he stopped. Right there in front of me, he died.

I held his hand, and closed my eyes, too. But, to be honest, I was relieved. I no longer had to worry about this poor boy, or his parents who didn't know that he was alive. And, I felt good knowing that I was there to comfort him when he needed it.

I was surprised that nobody had burst into the room. The bosses, some security, a medical team. Someone ought to be watching on all of these cameras. Or maybe they were all out to lunch, figuring that nothing interesting would be happening yet.

While I sat and waited, Carl's pale hands limply lying in my own, I thought about his claim that the mind works through some sort of arithmetic. That it's all just numbers underneath it all, just as comic book art is just dots.

I looked at the cabinets at the back of the room and thought about numbers. There were two cabinets, each had four drawers. That was simple. There were more numbers back there as well; the cabinets were twice as tall as the chair next to them. But what about opening a drawer? Where was the number in that? It was more of an operation than a quantity ... but of course numbers can represent operations as well. The number −2 decreases the values of other numbers when you add it to them. Additive inverses, like +2 and −2 could be like opening and closing the drawer.

Then, I started seeing it. Just as Carl had described, I saw the numbers underlying everything I was seeing or thinking. Not just quantities, but colors, textures, the hum of the fan ... they were all turning into numbers. And the algebraic properties of those numbers had a lot to do with how I perceived them. The carpet looked speckled because it had so many factors, the bright incandescent light was a high power of 2 and the fluorescent light was very close to, but frustratingly not quite equal, to such a number. Even my own thoughts, the logical arguments I was forming about what was happening to me, were arithmetic operations whose validity was analyzed through a modular arithmetic.

It was absolutely incredible, at first. As soon as a number popped into my head, I knew its number theoretic properties. I could see its prime factorization spread out before me ... not the way you might write it as a product of a bunch of integers, but visually. Large prime numbers are gorgeous, shimmering towers of ivory. The more factors the number has, the wider and more cumbersome it becomes.

And then, I lost the connection to reality. All I know now is the numbers. Carl described it well ... it is like I am *drowning* in numbers. I know that they must mean something. They might represent the black filing cabinets, or my bosses coming in to ask me to help them with crypts. It might be my own thoughts, my own heart beat. I don't know what they are. Some of them are beautiful numbers—Mersenne primes, perfect powers—others not, but they are all I have.

There are questions I will probably never know the answer to. Am I paranoid to wonder if this was their goal all along? Did my bosses send me in here to talk to Carl so that I could take over his job? Could they even have been involved in the accident that killed Brad and Liam? Well, I suppose that is not too likely. It is *somewhat* comforting, at least, to think that I'm now playing an important role in protecting my country, even if I'm doing it without being aware of it.

Is there something more to the cause of this "disease" than just thinking about it? Wouldn't it be ironic, for example, if some variant of the contagious virus that Brad and I predicted was responsible? Maybe I literally *caught* this from Carl.

In the end, though, the ultimate irony is that this is exactly what I always wanted from math. This is my mathematical aesthetic taken to the extreme. All of the complexities of the human world—suffering, art and music, politics, love—I now know that underneath it all is just a simple pattern, the rules of arithmetic that we learn in grade school. If that isn't beautiful, then I don't know what is.

5

Reality Conditions

I will proclaim to the world the deeds of Gilgamesh. This was the man to whom all things were known ... He was wise, he saw mysteries and knew secret things, he brought us a tale of the days before the flood. He went on a long journey, was weary, worn-out with labour, returning he rested, he engraved on a stone the whole story.

from *The Epic of Gilgamesh* as translated by N.K. Sandars

"Immortality" may be a silly word, but probably a mathematician has the best chance of whatever it may mean.

G.H. Hardy

Part I: The Factory

I can barely remember my past. No, that's not true; I remember the events and can recite them for you the way that I can tell you about the writing of the American Constitution. Just as I had trouble paying attention to history class because it seemed to have nothing to do with me, I feel as if my personal history is something I'm only remembering because I might be tested on it later.

I don't know why it is that my own life seems less real to me than television shows or books. If I happen to watch a few minutes of even the dumbest sit-com on TV, I find I have to watch it through to the end, to see what happens. Of course, I know the show isn't real, that someone just made it up to keep me interested enough that I'll put up with the commercials, but somehow I develop a belief in the characters that surpasses my belief in my own acquaintances and my own life. Only a few special memories stand out: the fire that destroyed the factory near my parents' house when I was young, Hank's death and the day at MSRI when my worldview collapsed. These things are relived when I remember them, as if they can be replayed with every emotion and sense responding precisely as they did when they really happened.

I don't know, maybe everyone feels this way. If not, if this is some neurosis of my own and not just part of the human condition, then it is at least partly my father's fault. I don't say that to put the blame on him and shift it away from myself. I have noticed that this is a popular cure these days: acknowledge your faults but blame them on your parents so you can feel justified in doing nothing to correct them. No, I am not angry at my father at all, not a bit. He was very comfortable living in his world and did his best to keep me there. It was my own attempt to get away, leaving me trapped halfway between his world and the one of ordinary people, that put me where I am today.

Father was a math professor, like I will soon be, and though I have developed some feelings of affection for his memory since he died, nobody would ever use the word "love" to describe what went on in our family. My father's world had few emotions. It was a world of pure intellect. Our house did not seem so much like a home, a place of comfort or for keeping your possessions, as it was a temple to the achievements of the human mind, where emotion and awareness of your immediate surroundings were to be avoided whenever possible.

I won't pretend to understand him and tell you I really know why my father was that way. One night while we lay curled up together on her livingroom rug, Maya asked me a lot of questions about him. As if her work as a bartender qualified her as a professional psychoanalyst, she announced decisively that my father's attitude had something to do with the Holocaust. She may have been right, but really I don't believe it. I don't think he was hiding from reality. I like to think that he really loved pure reason, that his avoidance of everything else was a sort of monogamy, a way of demonstrating his devotion. For instance, while all of the other professors we knew lived in nice homes in the sprawling suburbs, our broken-down house was halfway between the university and a nearby factory, one of the least desirable neighborhoods around. But who could doubt his devotion to the intellect when he was willing to live in such conditions just to be near the university?

That reminds me, I was going to tell you about the factory. From the time I was born until the month before my ninth birthday, this factory was running every minute of every day except Christmas. What it was making was scents, mostly the kind that they put in cleaning products like lemon or a field of wild flowers, but my favorite was the chocolate scent ... I have no idea what that one was for. I guess it must have taken them some work to set things up for making each scent, because once they started making one, lemon let's say, then they'd keep making it, vats and vats of it, for weeks at a time before switching to another. Then, for those weeks, my whole world (the house, my school, the playground) would smell like lemons. When the scent changed, from lemon to mint for example, the world changed with it. I was shocked when I realized that father paid absolutely no attention to which smell was being made. He had no favorites, no preferences. I remember thinking at that point that I did not want to grow up to be like him, so separated from the world that he can't even appreciate the things around him.

After the factory burned down—when I first heard the mysteriously whispered word "arson"—they didn't bother to rebuild, and from then on the world did not have a smell. There were smells: near the kitchen it smelled like whatever mom was making for dinner, near the garbage dump it smelled like garbage and so on. But the smells from the factory, that seemed to guide my life through a strange olfactory astrology, were gone. It was during those first few weeks after the fire that I first remember having this alienated feeling that has dominated my life.

By the time I was in high school, the sense of alienation was so much a part of who I was, I don't think I even wondered why I didn't have any friends. Everyone knew I was good at math, that I was going to be the valedictorian, and that was my identity. I didn't think of myself as one of the students in the school, as being even remotely the same as the other teenagers around me. My identity was unique: I was the school genius.

You might think that this would be a problem once I got to college. It is a shock for many high school geniuses when they get to the dorm at their Ivy League school and have to deal with the fact that they no longer stand out in a crowd. But I was saved from this dose of reality by my father. It's not that he prepared me for it in any way; I don't think he said anything to me about college other than "Okay" when I told him which school I was going to. No, my father shielded me from reality with his reputation.

Do you know any mathematics? If you do, there's a good chance you've heard my father's name: "Goldfarb's Series," "the Goldfarb cohomology," "Goldfarb Algebras." He made contributions to almost every subject in pure mathematics. The most amazing mathematicians, the ones who shake the foundations of mathematics every time they write a paper, they are called *gods* by the other mathematicians. Gunter Goldfarb was not a god, but he was royalty. He was a king.

Since I was a math major from day one, since nearly all of the people around me were math nerds, since I was royalty—Prince Gil Goldfarb whose very attitude showed that I was destined to be king—I still was unique, my singular identity was preserved, and I was saved from having to realize that I was part of the human race.

Don't get me wrong, I'm not claiming that I'm a mathematical genius. In fact, although I've met a lot of great mathematicians, I can't say that *any* of them is really like the mathematical geniuses that you see on TV and in movies. Some of them are really smart, smarter than me, but most successful mathematicians are just people who like math and put a great effort into it. That, more than any evidence of genius, was responsible for my success as a student of mathematics. I love it, and I worked at it whenever I could.

A lot of people are surprised that I went into math at all. Everyone knows that children are supposed to reject their parents' ideals, rebel. The truth is, I did. I focused my studies on mathematical *physics*, which definitely hurt my father. We never talked about it, not once. We both knew that my choice to do something so *applied* as this was a transgression against the temple of the intellect for which I could never be forgiven, but I like to think that I have received his forgiveness after his death.

My father died during my second year of graduate school. He wasn't so young that people viewed it as a tragedy, but he wasn't very old either. I know that I should probably have felt a terrible loss. My mother was overcome with grief and, apart from taking a few days away from my studies to attend the funeral, I was completely unaffected. If there is an afterlife and my father is looking down on me, then I think he would have been so proud of the way I completely ignored his death and my mom's pain, that he would have forgiven me for my area of research. At least, I like to think so.

Part II: Hank

In any case, though I was still in graduate school and had not yet proved even a single new theorem, I felt upon my return as if I had been crowned. Following my father's death, I was now king. Whether my classmates and professors knew this, I can't say. Though they had almost certainly read about my father's death in the *Notices*, nobody discussed it with me nor did they treat me differently. As before, I was never invited to social events. For anything from an afternoon at the University Pub to my officemate's wedding, I was not invited or even informed. On the other hand, I was always invited to seminars, thesis defenses and workshops. At this sort of gathering I had a special, unpopular role: I could always tell what it was that the speaker was hoping *not* to talk about. Whether it was a detail they had not yet fully figured out for themselves or something they just didn't know how to explain to us, I was able to recognize it and, by asking questions, force them to tell us about it anyway.

I don't think I had recognized that this was my role in the department, I was just going to the talks and asking questions when I had them. The fact that this was my own special niche, appreciated by the others because it was important even though unpleasant, was pointed out to me by Hank E. DuBois. Hank was my new officemate, taking over the desk next to mine after my old officemate got married and dropped out of school.

I had every reason to think I was going to get along as well with Hank as I did with everyone else, which is to say 'not well at all'. There were some superficial similarities between us: both about the same size and build, with similar facial features despite our different heritages, similar hair color, and—by a strange coincidence—he had lived for two years just on the *other* side of the factory from me when we were young boys. But in more important ways we were as different as we could be. While I was *always* completely clean shaven when I left the house, Hank had a shaggy, uneven beard, so sparse that you could see his chin through it even though each strand was quite long. Most importantly, Hank was remarkably well liked. He was always invited to social events, and often was the instigator of things like baseball games against other departments and trips to dance clubs. I never went to any of these things, but even I began to like him after we shared an office for only a few weeks. You couldn't help feeling, while talking to him about *anything*, that he was really interested in you and what you had to say. And he always had some comment to make that was either helpful or complimentary. I knew that this facade of politeness was probably just something he learned from his parents, just as I had learned to focus all of my attention inward from my father, but it was still very effective.

Hank and I spent a lot of time talking about my research while we sat there in the office. I was hoping to rework the mathematical foundations of quantum physics, improving on Von Neumann's rigorous but now somewhat archaic approach to the subject. It was my idea to reinterpret it in modern terms as a single eigenbundle over an appropriate moduli space of Riemann surfaces, with a single Hamiltonian operator acting on each fiber which has a dimension equal to the number of particles in the universe.

Hank was not so interested in physics, but still seemed to find the mathematical aspects of it enjoyable enough. I was really not used to talking to my fellow students about my research, having only discussed it before with my advisor and other senior researchers. So, I was surprised that he had as many helpful things to say as he did. Some of our discussions were over coffee at a cafe where you can sit on comfortable sofas amid collections of beat-up, old books. I completely ignored the books and did my best to ignore the soft cushions and bitter aroma of the coffee so that I could focus on his comments about my work, but one time, in the middle of a very important discussion, he noticed a book that he had been looking for lying on the table beside us. It was a copy of *The Blind Geometer* by Kim Stanley Robinson, a science-fiction story too long to be a short story and too short to be a novel. It was difficult to find, he told me, since it was published as a "double," stuck together with an unrelated story of the same length by another author. He said, "Gil, you've got to read this story. You'll like it! It has a lot of math in it." Despite the fact that I generally hate any kind of 'sci-fi,' I agreed to read it. It was not that I expected to like it, I knew I wouldn't, but I wanted to do it because Hank had asked me to!

I wasn't completely comfortable with the direction our relationship was going, but I had no way to stop it. His politeness was overpowering. After that, we saw some movies together, went to an art fair where his girlfriend was selling some of her sculptures, talked about our childhood experiences and were really *friends*. Even through all of this, we kept talking about math. He had become especially intrigued by one particular mathematical result in quantum mechanics, concerning the spectrum of Schrödinger operators. It was not something I was doing research on; it was old stuff from the late 1960's that I was just making use of in my thesis project, but he had a funny feeling about it. In particular, there was one paper from 1969 by David Kalman that contained a theorem on spectral decomposition that Hank felt had to be wrong. He looked into it a little further and realized that the statement of the theorem was not precisely true as it had been written, which in mathematics

is the same as being completely wrong. Hank not only had a counterexample to the theorem as it had been written, he also was able
to then prove a new, stronger and correct statement of the theorem.
This was very exciting to us—the first really important new thing
that either of us had ever done.

He wrote up a paper explaining Kalman's error and proving the
correct result and he showed it to his advisor. She saw right away
that he was right, and that this was important, but she suggested
that he make the portion in which he points out the error more
vague. "It's not a good idea to attack the gods," she said, referring
to Kalman's status as one of the most important mathematicians in
analysis today. Hank did tone it down quite a bit, after all it was the
polite thing to do, and in any case it would be clear to anyone who
read the paper that Kalman had made that mistake, so there was
no need to dwell on it.

He sent the paper off to *Bulletin des Sciences Mathématiques*, a top-
ranked math journal and received very favorable reports from the
referees who recognized that he had made an important contribution. When he got the acceptance letter, saying that the paper
would appear in the journal about one year later, we went out to
celebrate at the U Pub. As usual, I drank only iced tea, but Hank
got more than a little bit drunk. He was justifiably very proud of his
work, but his usual politeness had been washed away with the
drinks and he began to brag. Worse, as I helped him back to his
apartment, he began to say unkind things about some of our classmates. This was completely out of character for him, and I did not
like it. In fact, all of the bad things he said about them were completely true, but I was afraid to think of the things he might have
to say about me.

The next morning, he showed up in the office with a terrible
hangover. Despite this, he really wanted to talk to me about math.
He had another idea about my research, but he was feeling really
ill, so I told him to go home and I covered the two calculus sections
he was supposed to TA. That was the last time I ever saw him.

After teaching his classes, I graded the homework for a differential equations class, finished my own homework for my differential

geometry class and printed out two papers I had downloaded from the quant-ph preprint database. I had intended to read them at home, but instead, as soon as I lay on my bed, I ended up falling asleep with the papers on my chest. I overslept the next morning, not waking up until 9:30, but it didn't matter much because all of my appointments—a meeting with my advisor and some discussion sections for differential equations—were in the afternoon. So, I lazily made myself some coffee and turned on the computer.

The first e-mail message in my inbox was from my advisor, cancelling our meeting for the afternoon because he was too busy with some committee he was on. The second message was an announcement for that afternoon's speaker in the Geometric Analysis Seminar series. The third e-mail message waiting for me in my inbox was a general announcement sent to everyone in the math department from the secretary. She reported the bad news that Hank had been killed in an accident the day before at the Five Points. There were no details about the accident, just the news and his parents' address for sending condolences. I must have gone a little crazy at that point, because the first thing I did was disconnect the modem and try to call Hank at his apartment. Of course, there was no answer.

I was not thinking clearly, a mess of conflicting and overpowering emotions swirled around inside of me too quickly to be analyzed. Though I lied to myself and said that I was "just wandering aimlessly" when I found myself outside on the street, somehow I knew where I was going. It took a few extra minutes to get there because of intentional wrong turns to support the lie, but I very soon found myself at the intersection where the accident supposedly took place.

There was no sign of an accident that I could see. No blood splattered on the sidewalk. No broken windshield in the street. No chalk outline on the sidewalk. With no other thought of what I could or should do, my legs just collapsed under me, leaving me sitting lotus style with my back against a parking meter, comfortable but for the stench of old garbage and human waste from a nearby sewer.

Completely ignoring all of the strangers looking down at me with pity or derision, I focused my attention on a calming pattern, a mantra that fortuitously seemed to exist precisely for people in my pathetic situation. It was the pedestrian traffic signal across the street, repeating its preprogramed sequence of beckoning and warning. However, from where I sat, the words that it flashed were half blocked by a telephone pole. As a result, it comforted me with its endless chant of "Do Wa ... Do Wa ... Do Wa ... ..Waaaaaaaa ... Do Wa ... Do Wa ... Do Wa ... ..Waaaaaaa ..."

Part III: Maya

I seriously thought about attending the funeral, and probably would have if it had been here, but it was in Ohio where his parents live. Instead, on Friday, I went to the pub for my own mourning ritual. I was not sure exactly what this ritual was going to be. I really thought I might even drink something alcoholic, a going away toast to my one and only friend. In the end, I was sitting alone at the bar, drinking yet another iced tea, replaying in my mind the morning when I last saw Hank and trying to imagine what had happened to him. I heard from his girlfriend, who had heard from his parents, that he was crossing the street later that afternoon and had been hit by a minivan that sped through a red light. Where was he going? Did his hangover play a part in this? Had he figured out some important new connection between his discovery and my thesis project?

My bizarre behavior—sitting at the counter at the bar, looking solemn but not drinking—caught the attention of Maya, the bartender. Though I didn't know it at the time, she was a collector of odd people. Each of her friends was someone she treasured because they had some inexplicable eccentricity that endeared them to her, and she probably saw already that I qualified to join her circle of friends. At first, I resisted her requests to tell her about what was troubling me, but as I told her little pieces of the story, I started feeling better. Perhaps it was the way she responded to what I said, polite and interested, that reminded me of Hank. She certainly did not look like Hank! Petite, dark skinned and with a long straight nose that somehow fit perfectly on her little face, she really was quite beautiful.

I surprised myself by showing up at the pub again the following Friday, hoping to find her there again. Though I should have left as soon as I saw someone else behind the bar, I stayed and drank my iced tea. This time, I felt completely out of place, isolated and alone. When my drink was done, I got up to leave, but dropped my papers on the floor. Among the papers I was picking up were the two preprints I had downloaded the week before, still unread.

I snatched up the papers quickly, ran back to my office and sat at Hank's desk. His computer took a few minutes to boot up, and then I had to search through his files, but I eventually found the latest copy of his paper. If I didn't do this, then nobody would see his paper until it appeared in the journal in a year. It was suddenly important to me that his work be recognized immediately, as if the immortality that comes with a great scientific discovery had to be given quickly after death in order to effectively cure that disease. I uploaded his paper to all of the relevant lists at the arXiv preprint database: quantum physics, analysis, operator theory, differential equations, ...

Over the following weeks, I kept my eye on the preprints and the discussion groups, hoping to find some mention of Hank's result. I expected to see shocked researchers discussing Kalman's mistake; I thought I would see other researchers correcting earlier work to take into account the correct formulation of the spectral decomposition theorem; I thought Hank's name would be mentioned, not as a god or as royalty but at least as an important contributor to the theory, but there was nothing. Though the paper was available for downloading on the preprint server, I had no way of knowing whether anyone had done it. So, I wrote my own message to a few of the Usenet news groups asking about the paper. In order to get it posted, I had to make it look like an innocent inquiry:

```
To: sci.math.research
Posted-by: gilg@math.uruk.edu
Has anyone looked at
```

```
http://www.arxiv.org/math/0113218?
It seems to say some important new things
about spectral decomposition for
Schr\"odinger operators but I'm not
really an expert in that area. Can anyone
summarize for me how it fits in/alters
the field?
thanks,
Gil
```

There were only two replies. They both said essentially the same thing: Kalman fully addressed spectral decomposition in his 1969 paper.

I found this an unbearable thought, that Hank's work would go unrecognized, that his well-deserved immortality would be denied him while Kalman, already famous for his earlier work in another area, would still be remembered for his (incorrect) spectral decomposition theorem. I even went to speak to Hank's advisor about it, hoping to find in her an ally who would help me spread news of Hank's work, but she just said that she had expected this sort of thing to happen. She told me that there are a lot more mistakes made in mathematics than we realize. She said that even my father probably published a few mistakes. (I really *doubt* that.) Especially when the person making the mistake is a god, she said, mere mortals cannot hope to get much attention for pointing out the errors. She sounded to me as if she really thought that this was not only true, but just.

I stormed out of her office, slamming the door behind me. She probably thinks it was the remark about my father that made me so mad, but it wasn't. It was an anger at the world for taking away Hank's life and for not even rewarding him with appreciation for his only contribution to mathematics. And there was something else that made me run from her office: fear. Until that point, I never had any doubt that I, like my father, was going to be a mathematician whose name and work people know. I expected it because I thought I deserved it, but Hank deserved it more than I did. He had already done something, something I considered to be both important *and* somewhat brilliant. (I wish I had told him that when I had the chance.) If it will be denied to him, how could I be certain that such fame would come to me? What if I died before I proved *anything* important?

I walked off of the campus and started heading towards down-town, though I don't really know where I thought I was going. Someone called to me from across the street, "Hey! Hey math guy! You still down? We're going dancing. Want to come along?" It was Maya and a collection of her bizarre friends, three women and one man. I was wandering directionless anyway, and so I went with them. It was so much louder, brighter, and smokier than the U Pub; the dance club was unlike anything I had ever seen before. Once inside, Maya literally grabbed me by the arm and pulled me over to the bar. She said something to the bartender, but I couldn't hear what she said over the deafening noise of the band. "Try some of this," she shouted, handing me a short glass of milky liquid, "it's a Kahlua and Cream!" It wasn't bad, but I must have made a strange face because she laughed watching me drink it.

Then she dragged me out onto the dance floor. All around me were people jumping, shaking and moving as if their bodies were now controlled by the sound emanating from the throbbing speak-ers instead of by their own brains. On any other day in my life, I would have walked out of there and gone home to work on my math, but that day I stayed with Maya. At first, I felt awkward, expecting that the people around me were embarrassed to be stand-ing near someone who so obviously could not dance. Maya only looked at me occasionally, looking as often at her other friends, strangers or her feet. Then, when the next song began, somehow I was able to let go of my mind and felt my limbs jerking in rhythm. I did not feel like I was doing it, but it seemed to be working so I did not try to regain control. Maya nodded and winked at me to acknowledge the improvement. I was dancing.

Even though I'm telling you now that these things happened to me, I find it impossible to really believe myself. What was it that I saw in her? Probably just somebody who was willing to tell me what to do at a time when I was otherwise directionless. For the rest of that year, I was not myself at all. Although Maya complained about

how much I worked on my homework and my thesis project, I felt
as if I was barely devoting any time at all to those things that had
been the entire basis of my life. Instead, I was spending as much
time with Maya as I could. We both had strange schedules, me with
my classes and her with her bartending, but still managed to find a
way to be together almost all of the time. She took me to concerts,
to restaurants of every unusual ethnicity, to parties at her friends'
apartments and to her bedroom whenever there was nothing else
to do. Maybe you're wondering what she saw in me? At the time I
thought that she just thought that some of my quirks were cute.
She liked that my parents named me after a mythical Sumerian
superhero. She liked that in all that she showed me, from food to
sex, I was completely inexperienced and open to her suggestions. It
wasn't until later that I realized that I was also a challenge, a game
that she was playing.

You won't be surprised to hear that I was barely making any
progress on my thesis. I had noticed a few small interesting points I
could make, details about the mathematical foundations of quantum
mechanics that I don't think anyone else had yet observed, but nothing that would get me any real attention. As the semester drew to a
close, I was to have one final meeting with my advisor, Mike
Rosenberg, before the summer break. It had been my intention to
put in a great effort to have something stunning or exciting to show
him, but my classes and Maya took up more time than I had expected and so I showed up at his office at 5:30 with nothing new to say.
He didn't seem bothered by my lack of progress and said that it was
hard to get any important research done when one has responsibili-
ties and relationships to worry about. That was when Mike mentioned to me that he was going to be gone for the next semester,
attending a workshop at the Mathematical Sciences Research
Institute near Berkeley, California. He also invited me to join him
and gave me a flyer about the workshop he would be participating in.

I did not know much about MSRI. My father had gone there for
a workshop once when I was in college, but of course we did not
talk about it. I knew that people sometimes pronounced its
acroynm "misery," but always made it sound like it must be a nice

place to go. When I got down to my office, I looked over the flyer that Mike had given me. It listed the dates of the workshop, just a few weeks in October and not the entire semester at all, and the name of the workshop "Mathematical Aspects of Quantum Theory: A New Approach for the 21st Century." Then it listed all of the people who would be participating in the conference (though about half of them had *'s next to their name which meant that they 'had not yet confirmed their participation'). I got a sensation of vertigo merely from looking at that list of names. It was incredible! Every living mathematician *and* physicist I could imagine wanting to talk to about my thesis project was going to be there, gods and royalty alike.

I had grabbed my stuff and started walking home, fantasizing about how impressed all of these people at MSRI would be with my (nonexistent) results, when I remembered that I was supposed to be meeting Maya at Cafe Copa for dinner. At that moment, I could no longer think of *why* I had wanted to be with Maya at all. I was tempted to just blow her off, but turned around and went to meet her anyway.

She reacted badly to my suggestion that I would spend a semester in California. I don't think she was hurt, as you might think, by the idea that I wanted to be away from her. No, from the way she was talking about my research, I finally figured out what game she had been playing with me all along. I hadn't even realized! She was going to save me from math and turn me into a 'normal person'. Every time she got me to go somewhere and forget my work, she considered it a step towards victory. Of course, if I had ever really dropped math she would have won, but then would have thrown me aside. It had even looked as if she was winning and now suddenly, everything had changed. Out of fear of losing her game rather than any love of me, she begged me not to go. She told me how stupid the math research was, how it was just a way for me to hide from reality. I told her that math was more beautiful than any of the things she had shown me and that, besides, it was my ticket to immortality. Then I left.

Part IV: MSRI

I was just staring out the window, wondering about the lives of the people who I saw loading my luggage onto the plane, when a woman sat in the open seat next to mine. Once she was buckled in, she began tossing her yellow hair with her fingers like a salad of mixed greens, and each toss released a puff of a gingery aroma. She smiled at me apologetically, but did not stop until her hair lay just the way she wanted it. Then, taking a compact from the empty aisle seat on the other side of her, she applied a rouge to her already rosy cheeks. Again, she smiled at me.

Although she must have been at least twenty years older than me, there was something that I found very attractive about her. And so, I did not mind at all when she struck up a conversation with me just after take-off. We began with the usual polite small-talk, but eventually I told her that I was a mathematician on my way to a research workshop. I supposed it was likely that this would be a conversation stopper. Unfortunately, I was wrong.

"Oh, really?" she said. "You're a mathematician? Then you must have theories, don't you? That's such a coincidence, because I have theories too, though of course not professionally. Let me tell you ..." She put her hand on my thigh and leaned closer, until her ginger scented hair touched my cheek. The thing is, her theories were not mathematical theories, or physical theories like the ones I liked to consider. Her theories were *etymological* theories.

"For instance," she whispered as if telling me one of her deepest secrets, "I believe that the word 'butterfly' really used to be 'flutterby'. I mean, what do butterflies have to do with butter! But, they certainly flutter by, don't they? And then by mistake someone ... "

"That's an interesting idea," I admitted sincerely, "but what sort of *evidence* can you find for it."

"Evidence? Why, I thought you were a mathematician not a scientist or a lawyer!"

"I didn't mean *experimental* evidence, but mathematicians need to prove our theories. You can't expect anyone else to take your theories seriously just because you say so. I mean, have you ever seen an old document that uses the word 'flutterby'?"

"I really don't see why I should have to prove anything to anyone," she said calmly and confidently. "I just like my theory, and that's all. Don't you like it?"

What I liked was my theory, but I knew that I couldn't share it with anyone else until every lemma and theorem had been rigorously proved. I was annoyed at, and also a bit jealous of my neighbor's attitude.

Suddenly, she looked different to me, however. Her smile, instead of sexy, struck me as insipid. She was, I believed, the kind of person who would own a poster picturing a kitten hanging by its front claws from a tree branch, emblazoned with the maxim "Hang On". And this was not at all the kind of person I wanted to be with.

"Yes," I took her hand off of my leg and dropped it onto her lap, "it's very nice. I'm afraid I need to get some rest, though. Wake me up when we get there please!"

For a while, I only pretended to sleep. But only for a while.

Then, the stewardess woke me up. We had arrived in San Francisco and all of the other passengers had already left the plane. Although I was quite tired—it was already eleven at night according to my internal clock—I was excited and looking forward to my arrival at MSRI. I couldn't wait to meet some of the people who would be there for the workshop. I also was looking forward to seeing Mike, to tell him about the great progress I had made in my research over the summer. As if any was needed, this tremendous optimism I felt is final proof that I have no psychic ability to predict the future. For if I did, I would have had at least some inkling of how different this place was going to be from any other I knew, and how terrible I would feel by the end of my stay here.

I found the bus stop where I could catch the shuttle to Berkeley. When the bus—actually, more of a *van*—finally arrived, four of us got on board. Two others, a bohemian middle-aged couple, were bound for Berkeley like me, but a pretty young blonde woman was going to someplace called Emeryville. The driver said he would drop her off first and asked her to sit near the front with him. That made me think it was going to be a short ride, but it was not.

I was a little bit uncomfortable during the drive through the rolling hills south of San Francisco. This was my first time on the West Coast and I had not expected it to look so alien. During this time the driver was flirting with the woman going to Emeryville in the seat in front of me and the couple had fallen asleep behind me. The city was more familiar to me. Of course, the particular sky-scrapers looked different from those I knew in New York and Boston, but it felt like it could be home.

As we went over the Bay Bridge, a huge expanse of metal and concrete that connects San Francisco to Oakland, I could not understand why the blonde woman continued to flirt with the driv-er. Even if she found his tattoos more appealing than I did, I would have thought that his conversation would have revolted her. His idea of flirting was to talk about what a bitch his current lover was and how she had put out a restraining order on him when he said he was going to dump her! Moreover, his driving was really starting to scare me. He was holding the steering wheel only with his left hand so that he could gesticulate wildly with his right, but each motion of his free arm still had an effect on the vehicle. We would sway left and right, inadvertantly crossing over into other lanes, each time his arm swung. I looked down into the water below and at the huge steel pillars that we were passing at 70 mph and wished I really *could* escape from reality into my mathematics.

After we dropped her off in Emeryville, the driver actually left us in the van for two minutes so that he could run into a gas station convenience store for a cup of coffee. I hoped, in vain, that the cof-fee would make him a better driver. I was dropped off next at the apartment building where I would be staying (funded by Mike's grant). It was pretty much as the landlord had described it and would certainly be good enough for me to stay for the next five months. Best of all, there was a great little coffee shop around the corner on Walnut Street and a sushi restaurant less than a block away. (Sushi was the one thing that I kept from all of my experiences with Maya. It has a beautiful subtlety that I have become addicted to and in that way seems somehow to be very much like mathematics.)

The next morning I left the house early, expecting to get lost at least once on my way to MSRI, but I found the bus stop with no problem. Apart from an unusual species of tree here and there, I felt very much in my environment at the University of California campus. The people—professors and students—looked almost the same, the buildings looked almost the same. I was even able to identify Evans Hall where the math department is located from the back side (without seeing the name in huge letters on the front) merely by looking at the people who were going in.

The bus to MSRI leaves from in front of Evans Hall at UC Berkeley, even though MSRI is not technically part the school. When the bus, a very old little rusted van, finally arrived, I took my seat and tried to focus my thoughts on mathematics. Focusing on my work had been harder ever since Hank's death, but I was able to force myself to concentrate on it over the summer and had found what I expected would be my big break: a completely different approach to the Quantum Hall Effect. The Hall Effect, famous as one of the few well-known results named after a graduate student as much as for being a surprising macroscopic quantum phenomenon, would be the centerpiece of my paper. Providing a geometric explanation for this enigma of quantum physics would justify my whole approach. There were only a few lemmas I needed to prove in order to have the whole thing, and they seemed quite doable. So, I expected to be done with the project and ready to speak about it by the time the workshop was to begin.

But, as it turns out, I was not able to concentrate on the bus ride at all. Though I had looked at the bus route on a map, I was completely surprised by the fact that much of the way we seemed to be going up a very steep incline. MSRI was at the top of a *mountain* overlooking Berkeley and the Bay. Getting off of the bus was like entering a completely different world. While it was cloudy and completely overcast on the urban UC campus when I boarded the bus, I was now on a sunny mountain surrounded by a herd of long haired goats! The building that MSRI occupies did not look like any campus building I had ever seen. It is a rectangular building sided with a rough, unfinished, dark wood. In fact, it looks quite a bit more like a three story log cabin than I would have expected of a modern research facility.

I guess I was there too early, because the door was locked. So, I walked around to the other side of the building where I could see the view of Berkeley. As I walked on the dirt path that winds around the building, two small lizards scurried out of my way and when I got to the other side, I was completely amazed by what I saw. This was the view that greeted the gods of mathematics every morning when they arrived at MSRI, and nothing could have been more perfect.

I could not see Berkeley at all, although it was there beneath me, because it was entirely covered in still, white clouds. They completely surrounded this isolated hilltop, as if a tremendous flood of whipping cream had left the boxy, wooden Noah's Ark of a research institute here and now started to recede. The only things one could see rising above the clouds were the peaks of the Bay and Golden Gate Bridges, a few skyscrapers in San Francisco, and some mountains in the distance. I don't know how long I stared at this mythic panorama, but when I heard another bus arrive dropping off a large number of passengers, I went back around to the front door.

The door was still locked, but one of the newly arrived people was able to open it by waving her purse in front of a sensor at the right. Inside, a huge window revealed the same view I had seen from the path, but now Berkeley was beginning to appear slowly as the clouds were dispersed by the morning sun. The feeling of familiar comfort that I felt on campus was completely gone and replaced by uneasy awe. The central atrium with mathematically inspired sculpture, the piano, the absence of students and classrooms, these were all unavoidable clues that this was not just another math department.

After completing some paperwork, I was given an electronic key with which I could open the front door, a pass allowing me to ride the bus for free and a key to my office. My officemate had not yet arrived and so I was to have the office all to myself for the first two weeks. This would, I thought, give me the time I needed to prove the last few lemmas and get my ideas ready for the workshop. This seemed to be the ideal environment to do research and I was looking forward to getting it done.

In many ways, I had been right. The environment was conducive to research. I had nothing to think or worry about other than my work. The library, though non-circulating, had nearly all of the references I needed and any they did not have I could get from the physics library on campus. The view from my window was interesting and pretty, but fortunately not the hypnotizing view of the Bay that I saw the first morning. Instead it was a quiet view of wooded hills in Oakland.

The first two lemmas were proved before the end of the week, but the next three weeks were the worst of my life. Never before did I hate myself so much. I did not have the immediate gratification that comes from taking classes, learning new material and getting high grades on my homework. I did not enjoy the subtle feeling of superiority that I feel when teaching a class of undergraduates about mathematics that is beyond their ability but is child's play for me. I did not have Hank's friendship or Maya's physical intimacy. All there was in my world was this one, simple, stupid formula. The last lemma, entirely necessary for the rest of the work to mean anything at all that I thought I would be able to prove easily, was somehow evading me. Without it, I had essentially nothing but a huge list of conjectures. I *had* to prove this lemma before the workshop if I was going to get the respect I wanted from the other participants.

Each day I would stare at the formula, trying to figure out how I could prove that it was true. I wrote it in 36 different, but entirely equivalent forms. Each way, it was a surprising formula—your first guess would be that it is *not* true. But, I knew that it must be true. I had tested it in numerous examples in low-dimensional cases. Once the dimension got larger than three, however it became impossible to test any non-trivial examples. I also knew that it had to be true because it made sense if one viewed it physically and accepted my view of the rigged Hilbert space ... but that was not a mathematical proof. If I could figure out how to prove any one of these formulas, I would be done. But, every day I left feeling as if I had made no useful progress at all.

Actually, I should not have said that there was nothing in my world except for the formula. After the second week there was also my officemate, Utana. She was not part of the workshop; she had a postdoc position at MSRI simply to continue on her own research. I recognized her immediately since she was now quite well known. Her PhD thesis had gotten a lot of attention. Her name and picture appeared in all of the professional newsletters, and she'd already accepted a prestigious position at the University of Chicago to begin whenever she was ready. This seemed remarkably godlike for someone like her. It's not just that she was young. I'd met a young Field's medalist when he came to speak at our department once and not been as surprised as I was when I met Utana. I guess the surprising thing was how shy and unpretentious she was. She did not seem to have the *ego* necessary to be a famous mathematician.

Utana talked to me most mornings, but I was too absorbed in trying to prove my lemma to pay attention. She told me about the West African country she grew up in—I don't even remember now which one it was— and about her impressions of America. I generally only paid real attention to her comments about her success. She told me how lucky she was to have received a suggestion for such a good thesis topic from her advisor. (Her topic had some deep mathematics to it, computational difficulties that had previously seemed insurmountable, but also could be stated very simply so that even the *Times* and the science magazines could talk about it.) But, I learned, there were also some unusual political circumstances surrounding her success that I could never hope to reproduce. You see, she had obtained her degree at Albany University during the time that they were talking about eliminating the math department there. This had garnered a lot of press and generated a strong sentiment in the whole mathematical community. If Albany could fire all of their math professors (eliminating the department, it seemed, was a way one could fire even tenured professors) then no professors would feel safe. Moreover, if the view in Albany that math departments were not worth the money put into them was to become common then math professors would be especially at risk. So, when Utana's thesis seemed to be so marketable both inside and outside the mathematics community, there was an effort to make the most of it. First those within the department and then any mathematician who understood what was going on made sure that Utana's work was made to sound as important and as impressive as possible. After she had been covered in major newspapers and international magazines, after her work had been discussed as having important consequences, after her advisor got a large grant to run similar projects with other graduate students, it was impossible for the department to be eliminated. In the end, the administration backed down and said that they were no longer thinking of doing away with the math department or even the math graduate program. She had served her purpose, but now she was a star like no other recent PhD. Even the top graduates from Harvard and MIT did not have the name recognition that she had. Her immortality had been handed to her.

One morning, curious about the workshop even though she did not plan to participate, she asked me some questions about quantum physics. She was stunningly ignorant of the mathematical details or physical motivations of this theory. Apart from the nonsense you hear about in popular forums—like the supposed mystical implications of the Schrödinger's cat thought experiment—she knew almost nothing. Because she is a mathematician, I started by explaining the mathematics to her and was suitably impressed by how quickly she picked it up. She was not afraid to ask questions when she needed clarification. Some of the questions were just straightforward questions that mathematicians always ask, like why the underlying Hilbert space changes from example to example, but most of them were really very insightful inquiries that brought her right to the very points about the present mathematical formulation that trouble me. However, before I reached what I had expected to be the end of this "introductory lesson," she interrupted me with rapidly waving hands and a shake of her head.

"Okay", she said harshly, "I've got the math now, but what does this have to do with the real world. You call these eigenvalues 'measurements,' but most physical measurements I can think of in physics aren't so ... discrete."

"That's the illusion we have to be able to address, I guess," I said, though I could see that she would not be satisfied by this. So,

I tried explaining some basic experiments of quantum physics that would demonstrate the fundamental quantum nature of reality: the two-slit experiment, Bose-Einstein condensates and my personal favorite, the Quantum Hall effect. Again, the waving hands stopped me before I could finish.

"Did *you* ever do one of these experiments, Gil?" she asked.

"No, I ... "

"Maybe I'm too skeptical," she continued as if she had not expected me to answer at all,"but I have trouble believing in this bizarre sounding theory when it seems so contrary to what I experience every day. Is there any quantum physics you can show me in this room?"

It was difficult for me to see anything but quantum physics. What keeps you from falling through your chair? How are the electrons hitting the computer screen converted into light? How do your eyes detect that light in their photoreceptors? Finally, I hit on something that just looked quantized.

"How about this?" I said, picking up the cigarette lighter that was resting on the pack of cigarettes on her desk and twirling it between my thumb and forefinger.

"What? A little flint rubs on a piece of metal and lights a flammable gas! This is practically stone age technology."

"But look!" With a flick of my thumb, a small flame appeared that, despite Utana's claim, would certainly have amazed any stone age human. "Look at the flame. What colors do you see?"

"Most of it is sort of an orangish-yellow, and at the bottom there's just a bit of blue," she said cautiously, afraid that I was trying to trick her.

"Are there different shades of those two colors? I mean, does it get slowly more orange as it gets into the orange part? And what about where the flame *stops*? Does it slowly get less orange and eventually disappear?"

She looked more closely now, I saw in her expression that she was amazed never to have thought to look at this before. As she had said earlier, the classical physical measurements one usually thinks of all change slowly. Moving away from a hot flame, you expect the temperature to get gradually lower. In the language of mathematics one could say that she had imagined reality to be differentiable. "You're right," she said with sincere surprise, "all of the edges are sharp!"

"The soot at room temperature gives off no light and is practically transparent when rising through the air. But, when the temperature gets hot enough, it glows that yellowish color," I explained. "Then, when it reaches another threshhold it glows that blue color. There's no in between. It just looks like air, or it glows that yellow, or it's blue. It's quantized!"

She thanked me for my explanation, took back the lighter and turned back to her desk. At first, she went back to her research, but only for a minute. Then she pulled out the copy of the schedule for the Workshop on Quantum Theory that had been left in her box. When she noticed that I was still watching her she shrugged sheepishly and said "Maybe I'll go to a few talks after all."

Part V: The Workshop

I had grown to think of the lobby of MSRI as one of my retreats, a quiet and peaceful place where I could sit and think about my work while drinking a cup of coffee from the nearby kitchen. But, on the first morning of the workshop, the lobby was anything but quiet. A table had been set up with name tags for all of the participants. Looking at the tags that had not yet been picked up, and noting which had already been taken, I remembered and reexperienced the excitement I felt when I had first read the announcement for the workshop.

Wandering around this room, sipping from paper cups, talking to old friends, and introducing themselves to others they know only by reputation were all of the people whose work I have admired and emulated. The other graduate students and postdocs were standing in the corner by the piano talking to each other, afraid to mingle with the famous researchers in the rest of the room, but not me. I introduced myself to anyone there whose name I recognized: Yang, Berry, Chao, Kupershmidt. Some of them thought to ask whether I was my father's son and had nice things to say about him when I told them I was, but most just smiled politely and went back to whatever they were doing. One man, with a bright red sweater, grey hair and a thick Russian accent came over to introduce himself to me. "Hello," he said, "you are Gil Goldfarb, isn't it? Yes? Your advisor has told me about you. I am good looking very much to your talk."

I glanced at his nametag and found that he was Leonid Orlov from Moscow. "Oh, thank you, Professor Orlov, I am glad to finally be meeting you too."

"But now I cannot stay more," Orlov continued in his broken English. "I am meet my son who is in California on business. I will miss talks today. Back tomorrow."

"Oh," I asked, making small-talk while trying to remember why his name sounded so familiar, "what does your son do?"

"He is racist," Orlov answered.

"Yes, um ... but what does he do for a *living*?" I asked incredulously.

Orlov smiled. "It is as I say," he said. "Igor is racist *professional*! This is wonderful thing about America. If you are so good enough, you can get paid to do what you like!"

"But ... ?"

"Have you really never seen this, Gil? He has sponsors from beer, from cigarettes, and like so. They put advertisements right on his car."

"On his *car*?" I asked, starting to get it.

"Yes. My son is racist of *car*, not racist of horse. But, now I must go. I will see you later?"

Mike had not yet arrived, so I couldn't ask him who Orlov was. Instead I ducked out into the library and did a search of Orlov's publications on MathSci. It lists his most recent publications first. I could see that this guy is definitely a big shot, lots of publications in some of the most important journals in mathematical physics. All of his recent work, however, was on geometric quantization which is not something I pay much attention to. When I flipped down a few screens and saw his older work, I finally remembered where I knew him from.

Back when I had first gotten interested in working on QM stuff, Mike had suggested that I read a paper by Orlov that he said he thought I would really like. It was a paper from 1972, published in the Russian journal *Theoretical and Mathematical Physics*, on mathematical foundations of quantum mechanics. I did make an effort to read it, but did not get very far. At first I thought it must just have been the translation that I had, but I went and read the original in Russian and it was just as bad. There were claims, but no proofs and it seemed to have been written to make it difficult to read instead of to help the reader. I was not terribly excited by what he seemed to be doing, and I had the feeling that Mike had only suggested I look at it because Orlov was a friend of his, so I just put it aside and, as you can see, forgot about it.

Oh, well, Orlov seemed to have done some pretty impressive stuff more recently. So, I figured I would give his '72 paper another shot. Maybe this time I would understand it better. A paper that old would not be available at arXiv.org, so I pulled the journal off the shelf in the next room and made a photocopy of the article for myself. It was my intention to read it over as soon as I had time, but unfortunately that was too late.

The first talk was to be delivered by Rena Hagorn, a young hotshot in quantum gravity and former grad student of one of the workshop's organizers. It began at 9:15, fifteen minutes later than it was supposed to, but it was so fantastic that they let her run overtime by more than five minutes. It was a really great motivational talk, summarizing in a very beautiful way how our view of quantum mechanics has changed since its creation at the beginning of the twentieth century and making a good case for the need to completely recreate it on a new mathematical foundation. At the end of the talk, my mouth was watering and my heart was pounding with excitement. I felt as if I was part of history here, as if all of us in the lecture hall were going to work together to change the face of modern physics. This mood was completely destroyed by the next talk which was nothing but a dry, self-congratulatory review of some recent technical results that I did not find especially impressive.

And that is how the conference continued; there was an occasional talk that was unbelievably great, but mostly uninspiring advertisements by people I had expected to be better. Once Mike arrived on the second day, he would meet me and Leonid Orlov at the coffee breaks and we would critique each talk, generally unanimously agreeing upon which ones were good and which we should have skipped. But, even though there was agreement about which ones were good, there was a clear difference in the way Mike and Leonid responded to the bad talks. Mike's attitude seemed to be that everyone gives bad talks sometimes and so this did not necessarily mean much of anything. Leonid, on the other hand, viewed his analysis of the talks as irreversible decisions on the quality of the speaker and their entire body of work. For instance, "Okahawa is no good," he would say flatly. "His definitions are too sloppy ... does not even know Novikov's work ... no good."

One alleged purpose of the workshop was to unite mathematicians and physicists behind a common goal. In fact, it seemed to me that it emphasized our differences and pushed us apart. Of course, I had talked to physicists before, in one-on-one situations, but that was different. Here there were many mathematicians and many physicists, and though we knew we were supposed to be working together it seemed more like we were competing teams. For example, one Quebecois postdoc said to me "The work of mathematicians has not helped physics but physicists have had a tremendous impact on mathematics." I tried talking to him about this, to mention some cases that I thought were obvious counterexamples to his claim, but we ran into cultural differences. First, I wanted to know how he was defining 'mathematician' and 'physicist.' He thought this was a crazy question, but I thought we needed to address this first before we could go on. This is one of the big differences between physicists and mathematicians in general, the mathematician's obsessive interest in making certain that definitions are precise. For example, I thought that Isaac Newton was a perfect example: here was a mathematician who had practically *invented* physics, but my Canadian colleague said that Newton was a physicist. We eventually agreed that we could identify mathematicians and physicists by the field in which they got their degree. (This did not work for Newton—his degrees and jobs were all in mathematics but there really was no such thing as physics at the time.) So then I mentioned Sophus Lie and Galois whose algebraic techniques now form the basis of particle physics, the discovery of solitons by Kruskal and Zabusky, Riemann's notion of curvature without which one could probably not conceive of let alone describe general relativity, and I would have kept going but he interrupted me. He said "Yes, but physicists would have been able to discover those things on their own even if the mathematicians had not done it first." I personally do not believe that for a minute, and I don't see why I couldn't claim the same for our side, but it is beside the point. We had run into another cultural difference: physicists feel that they can change the rules in the middle of the game. For instance, they discuss in the same sentence a consequence of quantum mechanics and a consequence of general relativity without acknowledging that these two theories are completely inconsistent; you can't have both. Anyway, I'm sure that he is off complaining about me the same way. I'm not trying to make a judgement here, just pointing out the differences.

Part VI: Gil's Talk

Each evening, after a full day of listening to talks, I would go home and work on my own talk. I would have liked to show my latest results to Mike, but he was busy with a collaborator down on campus. Besides, he would say, he would get to hear me speak about it the following week when I gave my talk. Utana did let me practice the talk for her, and she gave me some good advice on how to present it, but since it was not her field she couldn't really critique the work.

Finally, the morning of my talk had arrived. I was to be the second speaker of the day. Although the first talk was one I would really have liked to hear, I surprised myself by being so nervous that I couldn't attend. Instead, I waited outside collecting my thoughts and my courage. I entered the lecture hall during the question session of the previous talk and stood by the wall holding my notes and some chalk until they were through.

Kuperschmidt introduced me, mentioning both my father and Mike even though my work is really not related to either of theirs, and quoting my title "Quantum Mechanics on the Rigged Hilbert Space: An Underlying Algebro-Geometric Structure and the Hall Effect". Like the other mathematicians who had spoken before me, I used the chalkboard and the first thing I wrote was a definition. (In contrast, all of the physicists projected their talks from a laptop and began them with either a spectacular diagram or a droll joke. This time I'm not just pointing out the difference; I think the physicist's way is probably better!)

After the first three definitions I was ready for a theorem, one of the minor ones I thought, and I only gave a hint of the proof. I was still quite nervous, and wondered if I was even speaking intelligibly. Looking around, I saw that some of the audience was barely paying attention, and a few were actually out in the hall talking, but when I looked down at Mike in the third row, he gave me an encouraging smile. I took it to mean that he liked the first theorem and after that I stopped being so self-conscious; the rest of the talk flowed out of me like a well rehearsed play. Using emotion in my voice as Utana had recommended, I was able to emphasize the parts of the talk I thought were the most important, and that seemed to work. The audience members who were not paying much attention in general paid attention to those parts of the talk at least. So, I thought, even if they didn't get the whole picture, they'd be able to remember a few isolated good results. There were occasional interruptions, people asking for clarification or pointing out connections to their own research, but I felt that I was doing pretty well time-wise; I still had ten minutes left when I reached my 'grand finale': explaining the Quantum Hall Effect in terms of the bispectrality of the Hamiltonian operator acting on the eigenbundle.

"Which do you mean," called a voice from the audience that I did not recognize, "the integer or fractional effect?"

"Both!" I said, unsure who asked the question and where to

direct my answer. "That's the amazing part. As you will see, they are both just consequences of the Riemann-Roch theorem in this set up." The tone of my voice (as well as the clock on the wall) made it clear to everyone who was listening that this was going to be the big finish. Even the people in the hall had stopped talking and were watching me through the doorway. As I was reaching the main result, I looked down at Mike, but my attention instead was drawn to Orlov sitting next to him. Orlov had a strange look on his face and was shaking his head slowly from side to side. I hoped this just meant that he couldn't believe how good the result was, but had a feeling that it was really going to be something bad. As soon as I had finished stating the main theorem, while I was writing the words "Summary and Conclusions" on the board, I heard Orlov calling to me.

"Gil," he said in a serious tone, "I'm sorry, but this is not Quantum Hall Effect. You know, I have this theorem in my paper from '72, but I did not need so much this algebra to prove it. It is quite similar to Hall Effect, except is no way stay in the reals. Number field is very important here; yours is complex numbers. So, nothing is quantized."

I did not understand what he meant when he said that my theorem was in his 1972 paper, and so I was really pretty worried at first, but I was relieved when I heard his specific objection. "No," I explained, "there are reality conditions here. The coherent states are chosen specifically to guarantee reality of all observables. I didn't mention that in my talk, there just wasn't enough time, but it is in my thesis and will be in the paper." Then, having addressed his concerns, I turned back to the board to continue with the wrap-up.

"No," Orlov called again, sounding angry now, "there is no reality condition consistent with theorem you state. I know this!"

I tried to think of what I could say now, but was saved by Kuperschmidt who said that we were running out of time. He suggested that I briefly summarize the key points and then we would break for questions, which is exactly what we did. There were a lot of questions, which is a good sign in general, but nobody asked anything about the main theorem, only a few of the minor details. After the usual polite applause, I left the lecture hall and ran up the stairs to my office.

When I reread Orlov's paper, I still could barely understand what he was saying most of the time, but now I recognized Theorem 3.6 as being equivalent to mine. It was stated in analytical terms, and of course there was no proof, but I could see that you could prove my "main result" using his Theorem 3.6 with almost no effort. So, he was right that he had done it before, but was he right that it could not be applied to explain the Quantum Hall Effect?

Part VII: Reality Conditions

The word 'number' is a more slippery term than most people would think. It isn't exactly clear what is and what is not a number, and the answers to those questions evolve in time. Historians of mathematics agree that at first the only numbers people considered were positive integers. Fractions, the result of dividing one of these integers by another one, were recognized and used, but they weren't considered numbers in themselves. It is a running theme here that what was once considered to be an *answer* to a question about numbers is later considered to be a number in itself. What do you get when you divide 5 by 8? We all know that it is the *number* $\frac{5}{8}$.

Similarly, negative numbers are an answer to the question of what do you get when you subtract a larger positive number from a smaller one.

If you think about it naively, the answer is that you *can't* do that; you can't start with 50 pennies and take away 100 of them. But, there is a good reason to consider negative numbers as well. To say $50 - 100 = -50$ is sensible if you think of each number representing not a quantity of pennies but the amount by which the total number of pennies *changes*. This is a subtle difference to be sure, but it is true that if you have enough pennies to start with, adding 50 pennies to the total number and then taking away 100 is the same as simply taking away 50 to start with, and that is what this expression means.

After hundreds of years in which positive and negative integers and fractions were considered to be numbers (these are what we now call the rational numbers) someone came along and proposed that even more things ought to be considered numbers. There were questions about numbers that could not be answered by any things that were considered to be numbers at that time. Here is an obvious one: if you take a string that is one foot long and make it into a perfect circle, how many feet long is the diameter of that circle? There is no rational number that answers this question, the answer is given by what we now call an *irrational number*. Believe me, there was a fight about considering these things to be numbers. You can't actually physically measure a circle and find precisely π anywhere using a ruler. Technically, I guess, π inches is a length on a ruler—unless you believe the loop gravity paper from a few years ago that claimed to prove that lengths are quantized—but you can't tell π from a very close rational number just by looking at it. However, in the end, the irrational numbers are considered numbers just like any other. Why is that? Two reasons, I guess: it is really quite useful to have such numbers because they make possible lots of important calculations and they follow the same rules of multiplication and addition as any other number, so why not consider them to be numbers? The extent to which we have taken this to heart is proved by the name we have given to signify all of the positive and negative, rational and irrational numbers. All together, they are the *real* numbers. There is no longer any question about whether the square root of two is a number just as the square root of four is a number; they are both real numbers.

The complex numbers are still in that in-between stage where lots of people, in fact almost everyone other than a minority of mathematicians, really don't think of them as numbers in the same way that the real numbers are numbers. Again, you can tell this from the name. Not 'complex number,' which sounds like it is a type of number even if it is a hard one, but 'imaginary number,' which is what you call a complex number with no real part. An *imaginary* number doesn't sound like a number at all. The name suggests that we want to *pretend* it is a number. Imaginary numbers aren't even on the number line, they are off on *another* number line that we draw intersecting the real number line at 0. So, why all of the interest in these numbers? As always, they are the answers to questions that we can't answer with any real numbers. What two numbers x solve the equation $x^2 = c$? The answer is always $x = \pm\sqrt{c}$, but if you want to know that the number x is a real number, you have to add a condition: c has to be *nonnegative*, or else x will be an imaginary number. This is a '*reality condition*.' It is a condition you have to add to make sure that the answer to your question does not involve complex numbers.

To many mathematicians, an answer that is a complex number is as good as any other, and why shouldn't it be? Complex numbers are numbers for all of the same reasons that the irrational numbers are now thought of as numbers. But, whenever somebody does a physics experiment and measures the mass, velocity, length, energy or position of something they always find a *real* number. Of course, this could be a limitation in our measuring devices, but that doesn't matter. If you're trying to explain the results of these experiments you have to end up with a real number. That is why it was important in my thesis to choose reality conditions way at the beginning. Orlov had claimed that my reality conditions did not work, that I was going to end up with complex-valued measurements no matter what I started with. So, I went back to my thesis and carefully looked over the section in which I proved, or thought I proved, that the reality conditions were valid. Upon rereading, it was clear that there was a hole in my argument. How could I have missed that? When I'm working, I do all of my calculations and write all of my thoughts down in bound notebooks, so I went back to my notebook from one and a half years ago to see what I had said about it at the time. It is very weird, but at the time I seemed to have been completely aware of the fact that my reality conditions were not quite right, but then I forgot about it. From some point on, I worked under the assumption that I proved it, but I never had.

I spent a little bit of time trying to see if I could fix the proof, to get some sort of reality conditions, but Orlov had said that it was not possible. He claimed to have shown, though it was certainly not in his paper, that any situation in which my theorem applied would by necessity involve complex numbers after a relatively short amount of time. I tried to talk to Orlov about it; I asked him to show me his proof, but he never really talked to me again. He would just excuse himself and say he had somewhere else he had to be, or he would pretend not to notice me. He may never have said it to my face, but in my mind I heard him saying "Goldfarb is no good ... doesn't even know my old work ... Goldfarb is no good."

Mike tried to be encouraging. He told me that people really liked all of the rest of my talk and that he had once been very proud of a piece of work that he later learned had been done earlier by someone else. This may all have been true, but I was not able to appreciate it. Somehow, for me, this had spoiled all of the fun. I had probably spent too long fantasizing about that moment when I was speaking to all of those people about my research and I couldn't deal with the fact that reality did not match my dream.

A few parts of my thesis had survived the removal of the reality conditions. I did not think they looked especially impressive, and at least one part—the one that bore a resemblence to the Quantum Hall Effect—had already been done by someone else. After the workshop ended, I did spend some time discussing these results that remained with other visitors and some professors from Berkeley. My confidence was gone, however, and I could not shake the feeling that they were talking to me only out of courtesy, or as a favor to Mike and my father. In any of these conversations, no matter how positively they ended, I would walk away hearing "Gil Goldfarb is no good" as if Leonid Orlov were whispering it into my ear.

The rest of my time at MSRI was used up by sending out job applications. Like all of the other finishing PhD's and postdocs I knew, I was sending out close to one hundred applications, to colleges and universities at all levels. The job market in recent years had been completely unpredictable. We had all been surprised to see who among the people that had graduated the year before got tons of great job offers and who only got one unappealing offer. So, I had no idea what to expect as a response to my applications, espe-

cially since the professors writing my letters of recommendation
had all heard by now about my big mistake.

Epilogue

I still think longingly of immortality from time to time. Gauss,
Riemann, Newton, Leibniz, Hilbert, Cartan. They have it, Hank
and I do not. We both came close, closer than most people. Hank
published a paper in a good journal, I gave a talk to all of the right
people at MSRI, but being close to immortality is no better than
not getting anywhere near it. Consider my new hero, Hannibal
Hamlin, as an example. You don't know who Hannibal Hamlin
was? No, he wasn't a mathematician ... at least not as far as I know.
But, you know who Andrew Johnson was, right? Hannibal Hamlin
was Abe Lincoln's first vice-president. He was Lincoln's vice-president until the *month* before the assasination. If he had only held on
for another month, he would have been President Hamlin and we
would all have known his name, but he didn't and so he is nobody.
Is it egotistical of me to think that I can identify with him?

It was very strange to return to the university after my semester
at MSRI. Everything seemed different. Of course, when *everything*
seems different, you know that it is *you* who has changed. Shortly
after I returned, I started getting responses to my job applications.
Mostly, of course, they were polite rejections. That was no surprise;
that seemed to happen to everyone. Eventually, however, I started
getting requests for interviews. Two requests came from major
research institutions, and a few from reasonable little colleges. One
came from one place that sounded so bad I decided not to even go
for an interview.

As it turns out, I did not get job offers from either of the big
research institutions. I did get offers from two of the colleges and
an offer of a one-year, non-renewable postdoc from a co-author of
Mike's in Vancouver. I haven't completely decided yet, but I think
I'm probably going to take one of the college jobs, a job as an assistant professor at an honors college in Florida. There are pros and
cons to each, but I'm leaning towards that job because I really liked

it there, because the students I met there seemed enthusiastic about mathematics and because my mom is in a retirement community not too far away. Of course, I won't have much opportunity there to work on my research, but I'm not planning to work on it for a while anyway. I still feel a pain—where, in my soul?—whenever I think about Leonid Orlov and my stupid mistake. If that wound ever heals, perhaps I'll be able to get back into it, or perhaps not.

You know, it's strange, but now that I've written this down, now that I can *read* the story of my life printed on a page, it somehow seems more real than it did when I was living it. I think I understand now the important difference between a story and reality. I had mistakenly thought that it was the presence of interesting things in stories, things too interesting to be part of real life, but now I see it is the opposite. It is the things that are left out of the story that give it power. Not just the details of daily life, every meal I ate and every time I did the laundry that are excluded from the story, but most importantly it is the author's decision of where to begin and where to end, leaving off all that precedes and follows. No matter what happens next, whether I take the job in Florida or not, whether I someday prove a really important theorem or not, whether I live or die, the real world will continue. But, because it seems like a good place for it, I can end this story here.

6

The Exception

Sam: Grandfather? Grandfather, are you awake?

Grandfather: Hmmm? Wha ... Oh, hi there Sam! I'm awake, I am. Just resting my eyes and remembering better days. Here, you move that stuff off the foot of the bed and have a seat. It's not everyday that I have a visitor.

Sam: Mom sent these cookies for you. Would you like one?

Grandfather: No, but put it in that drawer over there. The bottom one. If the nurses find it they'll probably take it away. That's very sweet of your mom. Thank her for me.

Sam: So, do you have a little time to talk to me about something?

Grandfather: Talk? That would be wonderful, Sam. What's the occasion?

Sam: Well ... .you know I like talking to you and all ... but I'm supposed to do a project for school. I'm supposed to *interview* you and find out about something you're really proud of from long ago.

Grandfather: Yeah, I remember doing a project just like that with *my* grandfather back in, oh, it must have been the 1980's.

Sam: Wow! I can't believe you were alive in the 1900's. It just seems ... so ... long ago, you know?

Grandfather: I do know. Sometimes it seems that way to me too, but other times it seems like it can't be more than a few years ago that I was your age.

Sam: And you had the same project to do for school?

Grandfather: Well, no. It wasn't quite the same project. My grandparents had a really hard time. We were supposed to interview them about how hard their lives were when they were young. How they were chased out of their country by people with guns, how they traveled to a new country that was very different, how they got through the depression and the wars ... .I should have been proud of them for all that, but they weren't proud of it. They just did what they had to do to survive. They did what they had to. For my generation, everything was easy. Except for a few battles, disasters and genocides that were no more real than TV shows to most of us, nothing really bad happened. We complained a lot, but surviving wasn't a problem. We didn't lack food or shelter or anything like that; we're just short on self-esteem. I guess that's why your project is different.

Sam: Uh oh, I didn't get all that. Is it okay with you if I record what you say for my report?

Grandfather: Record ... on what? Oh, is this that computer thing that your parents let them install in your head?

Sam: Sort of. It's not the computer that's in my head, just the Interface. I know you didn't like the idea, but it's *great*. I know that once this trial period is over, they'll make sure *every* school kid gets one.

Grandfather: Are you sure it will understand a gravelly old voice like mine?

Sam: Grandpa! Now come on, say what you said again, about your generation and that.

Grandfather: All right, ready? When I was young, I interviewed my grandpa and found out how he survived all kinds of terrible things that could have killed him. His generation really had to work together, to help each other. He didn't talk about pride at all. I think its funny that you kids are interviewing old people who survived a time where the big concern was whether we could figure out how to get disgustingly rich, and we're supposed to be proud.

Sam: You're not going to mess up my project, are you? Maybe I should go interview Grandma Aadra instead ....

Grandfather: Okay, okay. I'll tell you about something I'm proud of, and I bet you'll be surprised. You know, there are some things you don't know about me! What do you think I'll say?

Sam: You want me to guess? I bet you won't say something like marrying grandma or when dad was born; you aren't that gushy. It will be something about music I think.

Grandfather: You know that I played music?

Sam: Sure, grandpa, you were in a rock band. I've listened to some of the old files. It was kind of good in a weird way.

Grandfather: Ah, but do you know what I gave up to be a musician?

Sam: Gave up? No, I guess I don't know that one.

Grandfather: Before the band got big, I was a college student. Yeah, I went to the University of Illinois in Chicago, and I studied math.

Sam: Math?

Grandfather: Okay, I didn't really study math very much, but I was planning to get a degree in math. I just thought I could do it without studying.

Sam: And that's what you're proud of?

Grandfather: There's more, just wait. There was a professor there, he was an amazing guy. I mean, you could just look at this guy and see that he was a genius. It was like you could look in his eyes and see flashes from all of the brain activity. He was nice too; very quiet and friendly, not at all stuck up.

Sam: He was your math teacher?

Grandfather: He taught classes at UIC, but I never had him for a class. I got a summer job doing research with him. His research was, of course, way beyond me, but he gave me little things to do. Most of the things he gave me to do were things I bet he could have done in a minute but which took me a week or more, but he didn't rush me. It was part of some program to get students involved in research, so it was for me more than it was for him.

Anyway, ... oh, here's my favorite nurse!

Nurse: Hey Mr. Black. Hi, you visiting your grandfather? That's nice. Not enough kids do that kind of thing. Now, I'll just be a minute. Let me see your readings, please. There ...

Grandfather: Does it say I'm still alive?

Nurse: No, it doesn't, but we need to keep you here anyway so that we can keep getting money from your kids. Is that all right?

Grandfather: Sure, as long as I can get an extra serving of ice cream for dessert tomorrow.

Nurse: That's a deal, Mr. Black. I'll see what I can do. Oh, how about that? I must have read it wrong. It says you're doing well after all. Actually, these are the best readings I've gotten from you in months. It must be your visitor here. You know visitors are better medicine than the medicine! I'll let you get back to your company. Anything I can get for you before I go?

Grandfather: Yes, how about my *youth*? I seem to have misplaced it.

Nurse: Sure, I'll get you some down at the gift shop. What size are you?

Grandfather: Oh, get out of here. I'll see you tomorrow.

Nurse: You'll see me sooner than that. I'll be back in an hour with your Topoisomerex.

Sam: Wait. Did you know that a class action lawsuit was filed yesterday against the makers of Topoisomerex?

Nurse: Now how did you ... oh, is she one of those Interface kids? It must be wonderful to just see anything you want to know right there in your head. Now, did you look at the *whole* article or just the headline?

Sam: Well, I've got the whole thing now and ... oh. I see what you mean.

Grandfather: What?!

Sam: You're not pregnant, are you Grandpa?

Nurse: Then I guess we don't have to worry about the lawsuit, eh? I'll see you in an hour. Thanks for looking though, sweetheart.

Sam: She seems very nice, huh? Are all of the people here that nice?

Grandfather: No way. There's nobody else here half as nice as her. But, where was I in my story?

Sam: You were doing work with a math teacher. But I don't understand, what kind of "research" did he do?

Grandfather: Oh, there are lots of things in mathematics that people didn't know. I haven't looked lately, but I'm sure there is still a lot that they don't know. Some of it they *know* that they don't know and some they haven't even thought of yet.

Sam: Like what?

Grandfather: Well, I can tell you what we worked on. You know that if you multiply two whole numbers you get another whole number, like 4 × 2 = 8. Most numbers, like 16 and 12 you can get as a product in lots of different ways. You can get 16 as 2 × 8 and 4 × 4 and 16 × 1. Actually, you can write any number as itself times one. A *prime* number is a positive whole number that can't be written as a product of two positive whole numbers in any other way, just as itself times one.

Sam: Like 7, right? You can't get 7 except for 7 × 1.

Grandfather: Yeah, that's pretty good! You're not bad at math, huh?

Sam: Maybe. I learned what a prime number is, anyway. I don't even need to Interface to know that.

Grandfather: Okay, well here's a weird thing about primes and even numbers like 8 and 16 that you might not know. Someone noticed that whenever you look at an even number, it is the sum of two prime numbers. Let's try some, like 6 = 3 + 3, 8 = 5 + 3, 16 = 11 + 5 and so on. It's not true for every number, though. Some odd numbers like 11 can't do it. I mean, 11 is 10 + 1 and 9 + 2 and 8 + 3 and so on but you never get *both* numbers to be prime.

Sam: So what?

Grandfather: There's no 'what' ... I mean, this doesn't get you anything, it's just a question: Is it *always* true that an even number is the sum of two primes?

Sam: I thought you said it was.

Grandfather: Not quite, what I said is that whenever anyone *tried* an even number, it worked, but when I was a student nobody knew if it was always true. You can't just check every number, it would take forever. But, this brilliant guy I was working under figured out how to show that it was true for infinitely many even numbers.

Sam: I guess you're going to tell me how, right?

Grandfather: I'll try to give you an idea of it, but I can't say I honestly understand it either. Picture a donut.

Sam: What kind? I like honey dipped.

Grandfather: You wouldn't like this one ... it's all hairy.

Sam: What?!?

Grandfather: Imagine that on this donut at each point there is a hair coming straight out of it.

Sam: That's disgusting.

Grandfather: Nah, it's kind of beautiful. Each of these hairs is infinitely long; each one is like the number line.

Sam: Okay, a donut with a number line at each point ...

Grandfather: Exactly! If the number lines are all tied together, sort of glued together I guess, mathematicians call it a *line bundle*. And if instead of hair there are 2-dimensional spaces or 3-dimensional spaces or even bigger at each point then it's called a *vector bundle*. Don't try to picture these, it's impossible.

Sam: I'm looking up vector bundles. One of the good things about the Interface is supposed to be that it lets us "see" things in higher dimensions. Yup, there's a picture attached to the file. I think I can sort of see it, but I'm not really sure what I'm seeing.

Grandfather: That's okay, I wouldn't want you to understand it better than me anyway. The point is, that you can talk about one of these vector bundles on a donut where the dimension of the space at each point is some big even number $2n$.

This smart guy figured out a special kind of bundle, he called it a Goldbach bundle, which has a $2n$-dimensional space at every point, but because of the way the spaces are glued together, you can split the bundle apart into two vector bundles of prime dimensions. You see? You could start with a vector bundle that is 16-dimensional and split it into an 11-dimensional bundle and a 5-dimensional one glued together.

Sam: Oh, so after he did that then we can say we know it is true for every even number, right?

Grandfather: No, not quite. I didn't say that he could do this for every number n. His proof used something called the Riemann-Roch theorem and somebody's duality—some French guy I think—but that proof didn't work for all numbers. It didn't work if the dimension was $2n$ with n between ten to the eight thousand and twice ten to the eight thousand.

Sam: That's a lot of numbers.

Grandfather: Yes, but it's a lot better than they had before. Not only did it show that any even number big enough was a sum of two primes, it gave a *reason* why it should be true, and nobody had ever done that before.

Sam: What was the reason?

Grandfather: You got me! Something geometrical ... something like there must be enough *sections*. I think he thought he would be able to prove it like that for all numbers. He was looking for an elegant way to prove it, but he had to give me something to work on.

So, one day we met at a cafe in Greektown to talk about the project. He explained it to me, and I've just told you everything I understood from that, but then he asked me to help him. The only thing I could really do better than him was computer programming. You know, he grew up before there were computers, so he wasn't too good at them. So, he asked me to write a program to check all of the numbers that his proof hadn't worked for. There were a lot of numbers to check, but not infinitely many of them. Running on a really fast machine the program was done pretty quickly, and the result was a shocker!

Sam: A surprise? What, those numbers didn't work? I mean, you couldn't get those numbers as a sum of primes?

Grandfather: Most of them could. Out of all of those numbers, there was just *one* number, just one even number, which wasn't a sum of two primes.

Sam: That's weird! What's wrong with that one number?

Grandfather: I don't know! I don't think Professor Quotlbaum did either, but maybe they do now. All we know is that he checked infinitely many numbers and I checked the rest and only that one number didn't work. And that's what I'm proud of. I did that. I found that number. The one even number that can't be written as a sum of two primes and I found it. I knew about it before anyone … even before the professor!

Sam: But is it good for anything? Now that we know it, what can we do with it?

Grandfather: Okay, so it's probably a *stupid* thing to be proud of. Now that the problem is solved nobody cares about it anymore. But, still, I have to admit that I'm really glad it was me who did it! Mathematicians are like that. Oliver Hardy bragged in his book *A Mathematician's Apology* that he'd never done anything useful in his whole career. He was proud of doing pure, useless mathematics.

Sam: Actually, it was Godfrey H. Hardy who said that.

Grandfather: You … ?

Sam: Interface, remember. Cool, huh?

Grandfather: Kind of annoying, actually. But you get the point.

Sam: Well, according to what I just downloaded, it turns out that Hardy was wrong.

Grandfather: About what?

Sam: About his work being useless. The Hardy-Weinberg Equilibrium is used by genetic biologists all the time, and his number theory research is used by computers to ...

Grandfather: Okay, okay. I still wish your parents hadn't let them tinker with your brain. I swear, talking to you is like talking to an encyclopedia instead of a twelve-year-old girl!

Sam: Whatever. Anyway, if you like that kind of number stuff, why didn't you become a mathematician?

Grandfather: I wish I did. My roommate and I had a band, just the two of us, and one of our songs started getting popular. People all over the place were downloading it from our Website. We thought we were going to get rich and famous! That's what everyone wanted then, to be rich.

Sam: Rich? From music?

Grandfather: Yeah, it used to be a big money thing. You'd get a contract with a 'record label' and make millions of dollars, live the easy life. At least that's what we thought. Especially that year! The Supreme Court had just made a decision about internet music that we thought would kill it, but it turned around and bit them. People just stopped caring about the bands with contracts—their music cost too much. But I still thought there'd be money to make from people downloading music. Who would have thought that there were so many talented people willing to make music for free?

Sam: But your band was pretty popular.

Grandfather: We were, for another year or two, and we did make enough money for me to drop out of school without worrying about my future. But, when you come in and ask me what I'm proud of, I realized that it isn't the songs but the 'Black-Quotlbaum exception'.

Sam: Is that what they call it?

Grandfather: The 'Black-Quotlbaum exception' ... that's what they call the even number that doesn't follow 'the rule'. My name comes first because they do it alphabetically in mathem ...

Sam: I'll look it up.

Grandfather: First it's "Black", like our family name and then Q.. U.. O.. T.. L.. B.. A.. U.. M.

Sam: Oh yeah, I've got it. You guys wrote an article. Published in something called *International Mathematical Research Notices*. It's a really *popular* article, Grandpa.

Grandfather: Popular ... what do you mean?

Sam: A lot of hits, that's all. It's been downloaded by hundreds of thousands of people this year.

Grandfather: *This* year? That can't be right. Why, after all this time, would people still be looking at that old article?

Sam: I'll trace the links backwards and ... oh, there's a link to it from a news story a few months ago.

Grandfather: It can't be Quotlbaum's obituary or anything, he'd be way old by now.

Sam: Let's see ... Research Newsbriefs ... physics! ... blah blah ... elementary particles as eigenbundle ... recent results indicate that the dimension of this bundle and therefore the number of elementary particles in the universe is a multiple of the Black-Quotlbaum exception, the only even number which cannot be written as a sum of two primes! The number of particles in the *universe*? Wow! That is too weird!

Grandfather: The number of particles in the universe has something to do with *my* number? I don't understand how ...

Sam: All right, I'm definitely going to get an A on this report now!

Grandfather: Can you, you know, *print out* that article for me or something? I mean, did they install a printer in you when they wired you up? Particles in the universe ... maybe it wasn't such a stupid thing to have been proud of for all these years after all.

7

Pop Quiz

"Exactly where are you going?" he asked, both concerned and angered by his wife's sudden announcement that she would be out of town and unreachable for the next few days.

She put a stack of clothes into her suitcase and looked up at him. After looking deeply into his eyes for a moment, perhaps wishing that he could read her mind, she said "I'm afraid I can't tell you, though I really wish I could."

"Does this have anything to do with ... *him?*" He helped her to zip the bag shut.

"Him *who?* Do you think I'm having an affair? Tom, we've only been married for a month and a half!" Together they lifted the now very heavy suitcase and carried it to the door.

"No, that's not what I mean, I just can't remember the guy's name. That guy from Los Alamos that you worked with ... Mac something? You had to get a security clearance to work with those guys."

"Okay, I guess I can tell you this much. Yes, I got a call from Mac this morning. Something big has come up, real *big.* I'm excited, and I'm flattered they asked me, and although I hate all of this 'top secret' stuff, there is no way I'm saying 'no' to this."

"But why did they ask *you?* When you were working with them before, you thought that they didn't like your research, right?"

"It's not that they didn't like it, just that it wasn't applied enough. I work on projective algebraic geometry, just 'pure math'. There are some connections to things like waves and quantum physics that I, frankly, don't care much about. When they brought me in as a postdoc, they were hoping I could help them with some of that, but I couldn't."

"And so why the sudden change?"

"This is different, very different, and it looks like they need someone in my field, no connections to physics ... .Oh, look, do you promise to never tell anyone what I'm going to tell you now? They can't really expect me to keep secrets from my husband, can they? You'd better sit down."

He hated conversations that started with 'you'd better sit down' because that always means that what you are about to hear is going to make your head spin, something he never enjoyed. He sat down on the love-seat, expecting her to sit next to him, but she sat opposite him on the ottoman and took his hands in hers.

"Two months ago, some government station that watches out for this sort of thing found a new satellite orbiting the Earth. It really was new, like it wasn't there the month before, and they were pretty sure that nobody had put it there from Earth. In fact, they went back and looked at some images from Hubble from the previous weeks and they think they found an indication of its arrival." His head was starting to spin already and she had clearly not reached the main point yet; she held his hands more firmly to get his attention back. "Then they were able to get a close look at it somehow, I don't know if it was a telescope, another satellite or a space shuttle, but the conclusion was that it was actually an alien spaceship, probably unmanned—or unaliened or whatever—because it is so small! It just kept orbiting without doing anything, until earlier this week when it started sending out signals."

"Ohhhh," it sounded as if he'd been hit in the stomach, "and how exactly do you fit into this?"

"Are you kidding? There is always a lot of mathematics in these 'alien contact' things."

"You mean this has happened before?" he said, pulling his

hands away, feeling as if he just found out that he'd been lied to.

"No, no. I forget how you lose your sense of humor whenever a sudden change in plans comes up. I meant in books and stuff, like *Contact* and *His Master's Voice*. They always use math as the 'universal language' that we can use to communicate with aliens. Maybe the aliens have read some of my papers!" He had the feeling that she was waiting for him to laugh, but she must have been right about him losing his sense of humor because he didn't feel as if laughing was a possibility. She continued "Actually, the message the satellite is sending seems to have some mathematical content. I don't know exactly what, but it sounds as if it has something to do with projective algebraic geometry!"

"Sarah, what happens to the math experts in these books you've read?"

"Well, in *Contact*, Ellie was just publicly humiliated ..."

"Uh huh."

" ... and sometimes," she continued, still thinking she was being funny, "they start an intergallactic war or something. But, they always come home to their husbands, so don't worry!"

"I'm worried." He took her hands and bent his head down, as if to kiss them, but instead he just buried his face in them.

She was too excited to be really worried, but his words still echoed in the back of her mind later that evening as she was being shown to her room at the space center. Perhaps she should be worried. Her research never had any real world consequences, and there was safety in that. If she did something wrong here, or even something right, it was possible that something bad or even terrible could happen as a consequence.

She was sitting alone in the small, antiseptic bedroom, knowing that she was going to be 'called for' in about fifty minutes, and she started to worry.

When she was called into the meeting room, she was very comforted to see that the only person there was Franco, a mathematician she had known at the Los Alamos National Lab. Perhaps they knew that she never felt completely comfortable around the others and had thought of Franco as her only good friend there.

"Hey, Sarah," he said with a smile, "it's been a while. I wish I had time for small talk, but I think we have to assume that there's some urgency to this. Let me bring you up to pace as quick as I can. The satellite broadcast its signal as a standard Earth television signal. Since it seems unlikely that the aliens happen to use the same sort of signal as we developed here, we figured that the satellite has been monitoring signals from Earth to determine how to communicate with us. But, there was no Earth language used in the signal at all." With that, he popped a videotape into the VCP on the rolling cart and returned to his seat.

She saw right away why she had been contacted. The images were very geometric. First axes indicating a flat Euclidean space (as we call them here — who knows what alien they are named after) then an assortment of animated lines through the origin, each line collapsing to a point. Sarah got the idea, it was a projective space, the space you get by collapsing Euclidean space along each line through the origin to a point. This space was as familiar to her as her living room. So, when she saw the groups of dots that appeared at the bottom of the screen it did not take her long to figure out that they were homogeneous coordinates for a point in the space, even though they were written in base 13.

But, what were they trying to tell her? Were these coordinates of some point in outer-space that we need to know about? Their home world, maybe? Before she had much time to think about it, another collection of dots appeared below the first. It looked as if it could also be coordinates for a point in space, but one of the coordinates was missing. Instead of dots representing a number, there was an empty box there. Was the box also a symbol for a number? Perhaps it was some irrational number while the others were all integers?

"That's it," Franco said, pointing at the screen. "It stopped there and has just been rebroadcasting that image for the past week. Apparently, they are waiting for us to respond."

"What, is this a *quiz*?" she asked, but before she had even finished saying it she realized what the question was. There is more than one way to write the coordinates for a point in projective space, but there was only one way to fill in the box so that the second set of coordinates would correspond to the same point as the first set. "Oh, do they want us to fill in the box so that it is the same point?" She was pleased with herself for so quickly being able to figure out this alien message.

"That's kind of what I figured," said Franco. "If the box was replaced with 538, then both sets of homogeneous coordinates would be for the same point."

"You already know the answer? Then why did Mac send for me?"

"He didn't. I did, and I'm really glad you are here. I'm kind of worried about all this. This question was pretty easy, even for me, but what's next? And what happens if I get it *wrong*? I'd feel much better about this if you were working with me on it."

She understood exactly what he meant. She would have felt exactly the same way if she had been working on this alone, and she shared his fears, but hearing them from him gave her an opportunity to look at it differently. "Try to think positively, Franco. Maybe nothing happens if we get the questions wrong, but something great will happen when we get them right!"

Since they both agreed that the answer to the first question was '538,' they supposed it was time to send their response and see what happens. Franco went to get approval from whoever was in charge so that they could send a response to the satellite, but very carefully avoided calling this person by name.

It was decided that they should send the response, also as a television signal, using the same base 13 system involving sets of one to twelve dots for the digits. Sarah and Franco sat in a computer filled control center with the feed from the satellite displayed live on a large screen at the front of the room. There were several other people in the room with them monitoring the satellite, but only the computer in front of Franco contained their response. He hesitated before pressing the key that would broadcast their answer to the ship, looked at Sarah once again for approval, and then hit the key. As soon as the signal was sent, the 'question,' which had been sent continuously for more than a week, disappeared from the screen. It was replaced by another, similar geometric image. This time it was planes through the origin in a three-dimensional space that were animated and appeared to collapse down to points.

"I don't get it yet," said Franco. "Last time it was clear to me that they were talking about projective space, but what's this?"

"It's a *grassmannian*," Sarah said quietly, and then repeated more loudly "It's a grassmannian. Just like points in projective space are lines through the origin, points in a grassmannian are planes through the origin, or actually any subspace of a fixed dimension. Here it is planes through the origin."

While she was speaking, new dot-numbers appeared at the bottom of the screen. Franco pointed them out and said "Six coordinates, is that what you'd expect?"

"Not for the grassmannian they showed in the animation, but you can't really draw any interesting grassmannians. The simplest nontrivial one is $Gr(2; 4)$, the grassmannian of planes in 4-dimensional space, and it would have six Plücker coordinates." She grabbed a piece of paper and a pen from her backpack. First, she converted the numbers to base ten—it would have taken her too long to attempt any computations in base 13—and then she checked the coordinates in the Plücker relations. If these were actually Plücker coordinates then they would have to satisfy a quadratic polynomial.

"Yes," she said, drawing a huge check-mark on the page, "they are the coordinates for a point in $Gr(2; 4)$! So, what's the question?"

After a moment, another collection of six coordinates appeared on the screen. She verified that these too were the coordinates for a point in the grassmannian, though not the same point this time. Just as she finished the computation, a third set of coordinates appeared below the other two. Again, these were the coordinates of a point in the grassmannian. Finally, the question: a big empty box, large enough to hold the six Plücker coordinates for another point appeared at the very bottom of the screen.

"They want you to give them a point," Franco said, hoping to be helpful though he clearly did not know enough algebraic geometry to be able to do very much. "What do those three points have in common?"

Now she was starting to get worried again. This question was not as obvious as the first. It took her almost an hour, with Franco watching impatiently, to figure out what was going on. When she finally figured it out, she felt as if the lights in the room suddenly got brighter and the air suddenly became sweeter.

"It's not that those *three* points have something in common, it's just the first two. You see, grassmannians come with a natural *duality*, a natural way to associate a point in one grassmannian to a particular point in a certain other grassmannian. A point in $Gr(k; n)$ gives you a point in $Gr(n - k; n)$ ... they are isomorphic. In our situation, since $n - k$, that is four minus two, just happens to be two again, the other grassmannian is the same $Gr(2; 4)$ again. You see? Every point in $Gr(2; 4)$ has a *partner*, its dual, and the first two are dual!"

"Oh," he said excitedly, rejuvinated by the new understanding, "so they just want you to give the dual for the *other* point! Can you do that?"

"Sure," she said, grabbing a whole stack of papers from her bag, "that shouldn't be so bad."

It seemed to take forever for her to get an answer, though most of it was spent carefully checking to make absolutely certain that it was right. Again, after approval from 'the boss,' the answer was sent up to the satellite in its chosen vocabulary. Again, its only response was to send another question. This one was difficult even for Sarah to understand. This time, instead of Euclidean space, the initial image was of Riemann surfaces. There were elliptic curves and higher genus curves, and then again, lines on the surfaces meeting at a pair of common points collapse to points themselves. This was not a grassmannian, or anything she had heard of in algebraic geometry, but she had an idea of what they must mean. The coordinates sent this time meant nothing to her; she had no idea how to associate numbers to the geometric images she had just seen.

She excused herself and went to her room where she worked on understanding the last question, took a short nap, and wished she had not gotten involved in this. She had trouble making any sense out of this idea. It did not seem as if she could associate an algebrogeometric object to the idea that the video image suggested. When

Franco came to bring her some food almost 12 hours later, she was just starting to understand. She just needed to think about the singularities in a new way, so that some unpleasantly bizarre things were allowed to happen, but tamed by the fact that they could only occur at those isolated points. At last, she figured out what the numbers must mean: they were the coordinates of a pair of lines on two different surfaces of two different genuses.

Through the next day, she worked without sleep and with barely any food. Somehow, the circumstances allowed her to concentrate in a way she had never done before. She proved theorem after theorem about this new idea, when she finally hit upon the analogue of duality for grassmannians.

"In this theory," she explained to Franco, "a pair of lines on surfaces of genus g_1 and g_2 respectively would give us a unique line on a surface of genus $g_1 + g_2$ ... it must be the coordinates of that third line that they want!"

"And can you work out the coordinates of that line?"

"I already have!" She handed him a piece of paper with ten numbers written on it, both in decimal form and the aliens' preferred base 13.

"I'm sure about it, Franco. This theory is fantastic, it's beautiful. I've never felt so ... so ... so 'at one' with a new piece of math."

"Sarah, you must be very tired. Are you sure you are thinking clearly?"

He insisted that she explain every detail of the new theory to him, until he too was convinced that her ten numbers must be the answer to the latest question. It took him quite a long time to get approval from his superiors, but once again Sarah found herself in the control center ready to broadcast her answer. This time it was she who was to initiate the broadcast. Before doing it, she reconsidered whether she was sure her answer was correct—she was. She also thought about what she *hoped* would happen next, but she was not sure. Did she want to get *another* question? Did she want to be personally rewarded by the aliens? Did she just want the satellite to disappear without a trace?

All activity around her had stopped. The others in the room, rather than watching their monitors or the feed from the satellite, were all watching her. So, she took a deep breath and pressed the key.

This time, the mathematical images on the large screen disappeared, to be replaced by an image of a small metallic sphere in orbit above the Earth.

"What the hell is that?" came a voice over an intercom. "Did someone switch to the view from our observatory?"

"No, sir." replied the thin man sitting at the computer next to hers.

"As far as I can tell, this is still the feed coming directly from the satellite."

There was silence as the image showed a door at the bottom of the satellite opening up and an even smaller sphere falling from it and plummeting towards the Earth.

"Good god," came the voice over the intercom again, "can we track that?"

"I don't think so," said the man at the computer, "it looked awfully small to me."

Before anyone had time to say anything else, the monitors showed a horrific sight: a bright flash and rapidly growing cloud on the surface of the Earth near the location where the small sphere seemed to have come down. Then, suddenly, the Earth in the image on the screen was blown into hundreds of pieces by an explosion of unbelievable power. Several people in the room instinctively grabbed onto tables or chairs as if to stabilize themselves for the shock waves, but nothing in the room changed. The thin man at the computer clicked on an icon that opened a live view of the front gate to the facility. They saw sky, the security guard in his booth, a few cars driving past on the highway in the distance. Clearly, despite what they had seen, the Earth had not been destroyed. As the rubble and remains of the Earth in the image on the screen dissipated leaving only an image of empty space, the transmission from the satellite stopped abruptly. The screen now showed only static.

"What was *that* about?" Sarah asked rhetorically.

"The satellite is gone," said someone from behind her. They continued, pointing at their screen "It seems to have just flown away at high speed ... I think back in the direction it came from."

Sarah stayed at the facility for another two days, in case her special skills were needed again, but when nothing seemed to be happening they told her to prepare to go home. Before she left, Franco brought her some unpleasant news. It had been decided that everything Sarah learned at the space center would have to be considered classified. Of course, everything about the satellite had to be a secret, but she had been hoping that she could make use of the geometry she had learned while answering the third question. However, the word had come down from above that, at least until further notice, she was not to use or tell anyone about those ideas.

"I don't understand," she said, "why?!?"

"We really have no idea what happened here."

"Yes, Franco, I know that. But what harm could there be in my using those ideas? I really did figure them out myself, you know."

"I don't know, maybe they used those questions to *plant* those ideas in your mind knowing that they are somehow dangerous!" She stared at him to show that she knew that *he* knew that what he had just said was ridiculous. "All right," he continued, "I don't believe that for a second either. In any case, we have no choice. The decision was made *for* us."

"By whom?"

"That's something that *I'm* not allowed to talk about."

So, Sarah flew home and continued with her life, but she could not help thinking about her strange experience and her beautiful mathematical discoveries that she would never be able to talk about. Four months later, she had almost trained herself to avoid thinking about these things. Her husband had left for work and she was sitting down to breakfast when there was a knock at the door. Cautiously she looked through the peephole and was pleasantly surprised to see Franco looking very happy.

"Sarah," he said when she opened the door, "I got permission to come out here and show you this. It's good news!"

"Did something else happen?" She took a seat on the ottoman

and Franco sat on the loveseat.

"Well, a little over two months ago the satellite—or another one that looked like it—came back."

"And you didn't contact me?!?"

"If there had been any more geometry we would have. Instead, it just sat there again doing nothing just like it had the first time. Then, it started sending signals. This time there was no picture, just an audio track. At first we thought it was an alien language, it just kept talking and talking, but then we recognized that it was speaking in human languages. The first one must have been Chinese or something. So we contacted it and said 'We speak English.' Then ... can I play this somewhere?" He held up a video tape.

While she put the tape in the VCR, he continued, "It must have been learning all of the human languages, or at least the ones that broadcast television signals, while it was waiting. Okay, here it goes."

The screen was blank, not showing static, but simply black. A voice, sounding entirely human, said "You speak English?"

"Yes, and who are you?" said the voice that she had heard before over the intercom.

"That," continued the calm voice, "is something that we will not discuss in this conversation. I simply wanted to apologize to you for the misuse of my machine and of your time."

"You sent your machine here a few months ago? For what purpose."

"The machine has a simple purpose, it is sent far away and allows instantaneous transmission of signals between that distant location and another machine that I have here on my world."

"Instant communication? I don't think that makes sense, sir," said another voice on the tape that she guessed was not an alien. "Communication faster than the speed of light is not possible."

"If you think it is not possible to have instantaneous communication," the alien said in a friendly voice, "then you are mistaken. You are also mistaken to think that I had sent the machine to your planet earlier. It was not me, it was my child, an adolescent. You see, this machine is in common use in our society. It is used to retrieve information, to communicate with family and friends who are far away. It is generally acceptable to simply let our children use these machines, but recently there has been a misuse by many of our young, and I am afraid to say that my child is among them.

"School is very important to us, for many generations it has been considered the duty of the young to put all possible energy and effort into their training. But, along with a large number of misguided youth, my child had been avoiding his work by using the machine."

"You mean," said the human, trying not to laugh, "that was your child's *homework* that we were working on?"

"I'm afraid so. Our community has now recognized the problem and we are seeking to address it, first by apologizing to those like you who were tricked, and then by taking steps to make certain that this does not happen again. We apologize for any trouble that we may have caused." Then the tape ended.

"So what was the thing with the Earth blowing up?" Sarah asked.

"I guess it was done to distract us when the satellite left so that we wouldn't be able to observe how it moves or where it went. Or maybe it was just a joke; you know kids. Anyway, the good news is that although you still should not talk to anyone about the aliens, you are now encouraged to continue your research into the geometrical discoveries you've made." Franco smiled broadly, waiting for her to thank him.

"Why do I have the feeling that there is some part of this you still haven't told me?"

"If you insist," he said, "I'll give you the full message. They want you to continue your research, looking especially for any possibility of using it for a method of instantaneous communication across long distances."

She threw back her head and closed her eyes, finally understanding that they had not sent Franco here to reward her, or just to tell her about this because she might like to know. As usual, there was something very specific her government hoped to get from her. Still, she didn't care, she was looking forward to thinking again about these interesting questions. But, she wondered, would she feel differently about it now knowing that it is just some child's homework?

8

The Math Code

It was clear that something was wrong in Professor Coburn's office. The desk chair had been thrown back, knocking over the fish tank. The coat rack that was usually kept upright, behind the door, was now laying on the floor across the doorway. The crime scene, for that is what everyone presumed it was, had been discovered only an hour ago by a student who had come by for help with her homework. Now, a campus security officer and the head of the math department stood alone in the room.

"Isn't it possible," asked Prof. Lee, "that he just had an accident, like he slipped and knocked things over, and then had to run to the hospital or something?" The question she posed was a combination of her calm intelligence, which had led to her election as the department head less than one month ago, and panicked wishful thinking.

The officer looked slowly around the room, and said "I don't think so, professor. We already checked the hospitals and he isn't there. Also, you know I was a real cop before I got this job, and I've seen lots of crime scenes and lots of accidents. I can't be sure, but this looks to me like a crime scene." He walked over to the telephone that still sat, apparently undisturbed, on the corner of the desk. Using a piece of paper from the desk to avoid making any fingerprints, he lifted the receiver. "There's a dial tone, and the num-

ber to call in an emergency is printed right there on the phone."
He placed the handset back down gently. "If there was an emergency, why wouldn't he call?"

"Good point," agreed the professor, wondering for the first time
about the name of the man she was talking to and noticing the
name "BLATCH" on the tag above his badge.

"And what about this," he said with a surprised tone in his
voice, pointing towards the fallen coatrack. "There are no coats,
only a light sweater on the rack—after all, it is May—but here is a
knit skiing glove! Do you recognize this as belonging to Professor
Coburn?"

"No," she said, forgetting the seriousness of the situation and
getting carried away with the excitement of the discovery, "Alan
never wears anything like that. Even in winter he just goes around
in short sleeves. That's a great clue!"

"Actually," Blatch knelt down to get a closer look at the old
glove, "it is not a great clue. It would be hard to trace or get prints
from a ratty glove like this, but it means that we probably should
not expect to find the perpetrator's prints anywhere either."

They stood for a moment listening to the sound of the pump
from the fish tank, still plugged into the socket though it lay dry on
the floor. Officer Blatch pointed at the dead fish, most of which
were still near the fish tank though some had managed to get to the
other side of the desk before expiring. "I bet some crime lab could
examine those fish and tell us exactly when this thing happened.
Remember, don't touch anything, don't disturb anything."

"What happened?!?" The department secretary stood in the
hall, looking in over the disaster that was Professor Coburn's office.

"We don't know, Sofia," said Professor Lee. "Officer Blatch
here thinks that something bad has happened to Professor
Coburn."

"Yes, I'm afraid I'm going to have to assume that a serious crime
has occurred here. Either he is in very serious danger right now, or
maybe Professor Coburn is dead." Professor Lee stared angrily at
Blatch, clearly upset that he could suggest such an awful possibility. "I'm sorry. I think we have to be realistic here."

"You are probably right," agreed Professor Lee, "it is just hard for me to imagine how something like this could have happened."

Blatch turned away, looking out the window at the view of students crossing the quadrangle. "I can't help feeling," he said, "as if this was my fault. I mean, I was supposed to be patrolling this part of the campus when this must have happened. I ... "

"Don't blame yourself, officer. You can't be everywhere at once. Why don't we go to my office to discuss this further," she suggested "I need coffee ... and I need to sit down."

"That sounds like a very good idea."

As the three of them—Sofia, Professor Lee and Officer Blatch—walked around the hall to the main office, the officer stopped in front of the display case containing pictures of the department faculty. There was a picture of Professor Alan Coburn: more than a little heavy, but friendly looking, clean shaven and greying at the temples. There was a picture of Professor Xiaowa Lee, looking considerably younger in the portrait than the very somber woman beside him. He glanced over the rest of the pictures, nearly all of which looked the same since the cheap Polaroid camera made everyone look flattened and pale. The only one that stood out was the portrait of Professor Mike Rosenberg, which had been done with a better camera. This picture was apparently taken in a classroom during one of his lectures and showed him jumping off of the lectern, waving his arms and legs about like a madman.

When he was satisfied that he had seen enough, the three continued silently to the office where Sofia picked up the pile of mail and unlocked the door. Sofia followed her usual routine of first listening to the messages recorded on the answering machine and then sorting through the mail. It would take her at least 15 minutes to complete and, she hoped, would distract her from the unusual circumstances. This would be difficult since a large portion of the department faculty was beginning to gather around her desk, gossiping, crying and worrying about the very thing she was hoping to forget.

Meanwhile, in the next room, Officer Blatch asked Professor Lee whether she could think of anyone who might want to hurt Professor Coburn.

"No, of all of the people in the department he is the man least likely to have enemies. He is a truly brilliant mathematician, but unlike some of the others I have known he is also extremely likeable. Even his students like him."

"Does he have family, friends, *lovers?*" The officer emphasized 'lovers,' thinking that this would be a group especially likely to be related to a crime of this nature.

"Well, those of us in the department were really his only family and friends that I know of. And, as far as lovers go, I really doubt it."

"Then it's got to be money," Blatch said plainly. "It's got to be money."

Professor Lee's eyes opened wide. "The money ... oh my god! He did just win the Wolfson Prize in Algebra. It's worth almost half a million dollars. He was going to use most of it to start a new research center here, hire some postdocs, buy a powerful computer. Do you think whoever did this wanted to stop him from doing that?"

He couldn't help it; he laughed. "Most criminals don't care too much about math research, Professor Lee. No, I'm sure that whoever did this just wants the ... "

Sofia's scream cut him off before he could say 'money'. Blatch and Lee ran to the main office and found most of the faculty standing around Sofia's desk. Everyone stared at Sofia, holding the mail, too breathless to make another sound. Finally, she said just two words: "Ransom note."

Before anyone else could think of the implications, Officer Blatch grabbed a tissue from the box on the secretary's desk and used it to take the note from her hands. He laid it on the desk and everyone crowded in to see it. Most of it was printed, probably on a laser printer, but in the middle was a handwritten portion that was immediately recognized as some of Professor Coburn's remarkably neat penmanship.

This is what it said:

We have taken Professor Coburn. We read that his Wolfson Prize money "can be used for any purpose." Certainly it can be used to save his life. He is alive and well now. Professor Rosenberg will know that only Professor Coburn could have written this:

Yesterday in my office, I proved a new theorem. It was that for every pair L and B in Minkowski space, there is an A so that B times L tensor A is a subspace of Hilbert space.

We will kill him if we do not receive $457,892. Get the money together. More directions will follow.

After everyone had read it, Blatch asked "So, is Rosenberg here? Can he verify the statement?"

The faculty looked amongst themselves, with puzzled expressions. "He was here a minute ago." "Where did Mike go?" "I'll go find him. He can't have gone very far."

"Okay," Blatch said loudly trying to get everyone's attention, "while we wait for Rosenberg to get back, let me ask a couple of important questions." After a moment, everyone was silent and waiting attentively.

"First, is it really possible for you to get that kind of money?"

Many people mumbled an answer, but only Professor Lee spoke clearly. "Yes," she said, "just like they say. The money was given to him with no restrictions. It can be used for anything, and what could be a more important use than to save his life so that he can continue his research. I'm sure we can get the foundation to give us the money immediately if we tell them the circumstances."

"Good. Okay then," continued Blatch, "I should tell you that most police departments always tell people ... "

"Something is wrong!" interrupted Professor Calvino. "That mathematical statement in the ransom note makes no sense!" Once again, the math professors begin mumbling and speaking over each other.

"You're right, Francisco. I don't see how this can mean anything."

"Alan doesn't work with Hilbert spaces, why would he prove something about Hilbert spaces?"

"Which Hilbert space are we talking about? What is A?"

"This must mean that they *don't* have Alan ... or that he isn't alive!"

"Hold on," said Professor Lee in the firm but calm voice that always quiets down a crowd. "Why would the kidnappers put in something that doesn't make sense? It certainly isn't going to fool Mike when he gets back, and the note specifically says we should show it to him. That wouldn't make any sense. I think the only sensible thing is to think that Alan *did* write that, and that the kidnappers did not realize that it was nonsense."

"But then, Alan must not be thinking clearly. He is not well! They hit him on the head ..." Calvino said with his finger in the air.

"It doesn't matter," said Blatch. "It doesn't matter. Like Professor Lee said, we still have to assume that they have Professor Coburn as a prisoner. As I was saying before, police departments always tell people in this situation that they should just do what the kidnappers ask. I mean, don't start thinking that this is a movie and there is going to be a dramatic rescue and a happy ending. Unfortunately, in reality, these things only work out nicely when the bad guys get what they want."

"That's correct, though not quite the way I would have said it, Blatch." Officer Blatch turned around to see Captain Miller, the head of campus security, standing in the doorway. "Professor Rosenberg has explained the situation to me, and I think I'd better take over operations at this point. We've talked before about the chain of command around here, Blatch, and we'll talk about it again later."

"Yes, sir," said Blatch with as little respect in his voice as possible, "I'd better get back on the beat then."

"Captain, I hope you are not going to take out your anger on Officer Blatch." Professor Lee insisted, "This is a tough situation, but I think he's been doing a very good job."

"Thank you for that input, ma'am," said the captain, "you are probably right. Blatch, why don't you stay here with me and help me out? Perhaps you can show me how its done!" Blatch, who had reached the doorway, turned around slowly and came back to the desk.

"Mike," Professor Lee called to Professor Rosenberg, "take a look at this ransom note. It mentions you! Does this 'theorem' mean anything to you?"

Instead of walking over to the desk, Professor Rosenberg walked to the whiteboard behind the secretary's desk and began to erase it. Sofia tried to get his attention by waving her hand, to indicate that he should not erase those important memos, but it was too late. "Actually," he said, "I read the note earlier and that's why I left to get Captain Miller."

"But it is meaningless nonsense!" shouted Calvino. "What is this 'B times L tensor A is a subspace of Hilbert space' stuff?"

"Mathematically this has no meaning, but we can still write it down in mathematical notation," Rosenberg explained, "and that's where we'll find the message that Alan is really sending us."

"You mean *this*?" Calvino came up to the whiteboard, uncapped a pen and wrote:

$$BL \otimes A \subset H.$$

There was nodding and mumbling among the professors crowding the office. "Yes, I guess that's how I'd write it." "I still don't get it." "What is the tensor *over*, complex scalars?" At this point, Blatch again turned to leave the office, but stopped when he noticed two other officers were standing there, blocking the door with their hands on their guns. He knew that there were three other officers on duty at this time as well and wondered where they might be right now.

"That's right," said Rosenberg, "except that when L and A are both vectors—as they seem to be here—you can write the tensor product as ... "

Calvino erased the symbol $\otimes$ from his expression and added a T for 'transpose' atop the A which left him with

$$BLA^T \subset H.$$

Everyone turned to look at Blatch who was sweating and blinking rapidly.

"I *still* don't get it!" said a professor somewhere in the crowd. A whispered voice explained it to him and when he finally understood he said "Ooooh!"

"Oh come on, that doesn't prove that I did anything," Blatch implored.

"True," Rosenberg agreed, "I could probably have written it a different way. But this is the only way I can think of that contains any meaning. Not *mathematical* meaning, of course, but meaning of another sort. You see, Alan is using mathematical notation itself as a code. And besides, I'm hoping that pretty soon we really *will* have proof."

Blatch was going to ask him to explain, but before he could the captain's radio started making awful clicking noises. It sounded as if the radio were broken, but the officers all knew that this just showed that someone was trying to reach him. "Miller," the captain said into the microphone attached to his lapel.

Through the static, everyone in the room could clearly hear a voice say "We found him, captain, and he's okay. He's fine." The captain's response was drowned out by cheers and sighs of relief from the gathered faculty. When that was done, however, more than a few people—including Blatch—turned to Professor Rosenberg since he clearly had not completely finished his explanation.

"You have to write out the *entire* statement of the 'theorem', not just the result but the conditions as well. Mathematicians make lots of statements about 'every' something, so we abbreviate 'for every' by the symbol '∀'. (I guess it is an upside-down 'A' for 'all')."

"Oh!" Calvino said, adding new symbols to the expression already on the board. "So 'for every L and B in Minkowski space..'"

$$\forall \, L, B \in \mathbb{M}$$

"' ... there exists an A such that ... '"

$$\exists \, A \text{ s.t.}$$

"Now *I* don't get it! What does this mean?"

"Well, what is Minkowski space?" Rosenberg asked, as if he was teaching a class.

"It's four-dimensional Euclidean space, but with a different metric, *time* has a negative signature like in relativity," answered someone.

"That's right! So, as long as you remember to use the product given by that metric, you could just write $\mathbb{R}^4$ instead of $\mathbb{M}$." Now

Rosenberg took the pen and corrected the senseless mathematical expression on the board to say

$$\forall L, B \in \mathbb{R}^4 \; \exists \, A \; s.t.$$
$$BLA^\mathsf{T} \subset H.$$

"I'm not sure I understand what you guys are all talking about," said the captain, "but I can tell you that we found Professor Coburn locked in a closet exactly where Rosenberg said he'd be, an empty house at 4 East Alber Street."

9

Monster

To Make Life Better

"... and this morning's broadcast of your local news is brought to you by JAMcorp. JAMcorp, using the power of mathematics to make life better for everyone."

[cheerful jingle:] From your morning coffee to your evening bath, we're improving it all with the power of (click)

Switching off her car radio, Quasia Fine curses to herself. "Damn annoying JAMcorp," she says. "That crap they do can't be called math. Preventing ketchup bottles from sounding flatulent is not math."

According to the digital clock in the dashboard, it is now 9:24. The fact that this is incorrect—it is only 9:21—would be of no consolation.

Even in the highly unlikely event that there is no traffic on the FDR Drive, there is no way that she could get to work on time now. Quasia hates being late, which in part explains why she is ranting to herself about ketchup bottles, but—truth be told—it is mostly because she hates her job as a math professor.

The funny thing is, she can't figure out why she hates it. It is exactly what she always wanted to be.

"Oh no," Quasia says aloud, "here we go again." Bracing herself for another dose of JAMcorp's cheeriness, she decides to try to outsmart it. The exit lane she is in will take her off of the Triboro Bridge and onto a twisted ramp that winds around the skyline of Manhattan like a little roller coaster. This is the perfect circumstance for JAMcorp's first product: "Groovy". These pavement grooves are the sophisticated progeny of the groves that used to precede toll booths and unexpected stop signs, but those primitive grooves merely caused the car to vibrate in an annoying hum. JAMcorp's first improved version varied the pattern of the grooves so that they make the very frame of the car recite a brief recorded message. This most certainly involved some very clever use of mathematics to get frequencies other than the natural harmonics that the frame of the car would generally produce. (Some experts in asymptotic p-adic analysis, in fact, have claimed that the engineers at JAMcorp must have made use of some theorems in that field that have not yet been announced publicly.) At first, "Groovy" was made available to government highway agencies for free, so that ramps such as the one she was rapidly approaching could say something less than thrilling like "Drive Slowly", but that was only a brilliant way for the company to beta test and advertise their new product. The new "Groovy" came out shortly afterward, and as well as working out a few bugs (bad reactions with certain suspensions or people driving at unusual speeds) it also greatly expanded the possible frequencies so that a brief jingle could be played ... and of course, new improved "Groovy" was no longer free. Faster than anyone could imagine, musical "Groovy" messages were showing up at highway exits advertising restaurants, hotels, tattoo parlors, and more. More recently, even the governmental "Groovies" have become musical and, in her opinion, utterly too cute. So, as she comes around the turn, Quasia's car vibrates so as to sound just like Simon and Garfunkel singing 'Slow down, you move too fast ... '.

Quasia had tried her best to mess up the song by driving fast and skidding a bit on the turn, but to her frustration, it sounded pretty good anyway.

The Dream

Quasia Fine had an unusual dream for an adolescent girl. No one she knew, not her parents or her sister or her friends, understood why she wanted to be a math professor. But this lack of support did not deter her one bit. For most of her teenage years, the dream remained constant and clear. This was the dream:

"Good morning, Professor Fine," said her handsome but nerdy colleague in the chalk covered tweed coat.

"Good morning, Professor Schmidt," she would reply. "And how are you today?"

Schmidt pushed his glasses up the bridge of his nose with the back of his hand. "Just fine, just fine. Last night I finally proved the Infinity mod Zero Conjecture!" he beamed, oblivious of the chalk mark he'd just made over his left eyebrow.

"Congratulations, Graham!" she said sincerely, knowing how hard he had been working on that one. "That will surely get you a renewal of your grant, and probably put you in line for the Kovalevskaya Chair of Mathematics."

"Oh, come on. It wasn't anything, really. Just a proof by contradiction using induction and a bit of chaos theory. Anyway, how are you, Quasia?"

"I'm doing rather well myself, Graham. This morning on the ride in to work ... "

Just from the gleam in her eye, Professor Schmidt guessed what she was going to say. "You didn't ... did you?!?"

"I did, Graham, I really did it. It was your advice yesterday about the fractal dimension of the solution space that was the last straw. Then, it all came together for me this morning. I wrote the details down in my office just to be sure, and I am."

"This extends Fine's Theor ... I mean your theorem to every imaginable space. The Fields Medal committee can't help but take notice of you now, Quasia. Nobody could."

When humility forced Quasia to look at her feet, a few strands of her thin brown hair fell across her face and Graham Schmidt gently guided them back into place, finally leaving his hand resting on her bare shoulder.

"*Quasia, you are so brilliant, I was wondering whether you could meet me for dinner tonight to ... um ... help me out with a project I'm working on?*"

"*With what, pray tell?*"

"*It's a problem of dynamics, I suppose ... I'm interested in minimizing the distance between two bodies.*" Now, his strong fingers were curled around the back of her neck, his thumb slowly massaging her earlobe.

"*Ah, I see. Is there a thermodynamic structure?*"

"*Yes, I'd say that there was definitely heat involved. I'm hoping that it could eventually lead to some ... intertwining relations.*"

"*Oh, and would these intertwined operators be bounded or unbounded?*" Quasia asked shyly.

"*That's a surprising question, Quasia—but an interesting one. I hadn't dared speculate about such details yet, the whole thing is just a conjecture.*"

"*You know, I once tried to prove a similar convergence theorem myself last year and got nowhere. You don't suppose this is somehow a corollary of my new theorem?*"

"*Oh no, Quasia. You know, I've been working on this two-body project for years now, and I think I may finally be getting close to ... *"

"*Professor Fine, I'm sorry to bother you, but it's ten o'clock,*" said the student in the classroom doorway. Through the door Quasia could see all of the students at their desks, impatiently waiting for her to begin her lecture. "*We're really looking forward to the rest of your lecture on Goldbach's Conjecture.*"

"*Very well,*" she sighed. Entering the classroom, she was startled by the applause and cheers from the rows of students. Turning back to look at Professor Schmidt still staring at her from the hallway, Quasia shrugged. He just smiled, winked, and walked away.

Every step of the way, it looked as if she was heading straight for that dream, doing everything that could be done to make it a reality. As an undergraduate, she impressed her professors not only with her broad knowledge of advanced mathematics, but also with her abilities to quickly learn, completely understand and explicitly prove difficult theorems. Of course, they wrote her glowing letters of recommendation that got her into one of the best graduate schools working under one of the most famous advisors. Her thesis, though not Field's Medal material on its own, was definitely an important contribution to representation theory and got her a job as an assistant professor straight out of graduate school while most of her fellow students had to be content with temporary positions and 'postdocs.'

Reality Bites

Now a tenured professor, Quasia no longer has the naive views of mathematics or academia that attracted her as a teen. Still, in the back of her mind, she could not help but wonder where she went wrong. Why did her life seem so completely different from her dream so as to leave her feeling so unfulfilled?

This question was usually buried beneath a writhing pile of things that she had to think about more urgently: preparing her lectures, buying a gift for her in-laws' golden wedding anniversary, writing the *Math Reviews* that were already two weeks overdue, etc. However, during especially bad situations, it would rise to the surface and beg her frontal lobe for an answer.

Right now is one such occasion. Right now, Quasia is ten minutes late for the third meeting this week of the faculty committee on committee assignments. This is the third meeting because her committee is stuck in a deadlock, with three members supporting each of the two proposals and the seventh refusing to support either. (Interestingly, the topic they are supposed to be addressing is a proposal on how to handle the deadlocks that have recently been plaguing the faculty committees.) Even though she is ten minutes late, she has not missed anything. They had decided to wait for her; after all, she is the chair of the committee.

Being chair forces her to actually pay some attention to what is being said: Pete Moss from the philosophy department calmly presents his argument; Ida Claire from history heatedly points out serious problems with Pete's suggestions; the bald guy from the physics department (Howard? Hubert?) points out that all of the proposals presented so far are unfair to departments from the School of

Sciences. However, Quasia finds that her general lack of interest in the topic allows her to continue her duties as chair while letting the rest of her mind consider the question of how her dreary reality differs from her adolescent fantasy.

One obvious difference is that, unlike in her dream, real professors never call each other "Professor" ... except as a bad joke. But surely this could not account for her unhappiness: a rose by any other name and all that.

A part of the dream she had consciously given up on very early in her career was her magical fantasy romance with "Professor Graham Schmidt." Her experiences with dating during her years in college had taught her enough about romance to blow away that aspect of the dream.

All of her relationships had been okay, but far from magical, and much more trouble than they were worth. Then, during her second year of grad school, she had a relationship that was surprisingly wonderful. His name was Hugh Hu. He had the nerdiness of the imaginary Professor Schmidt without his wit or charm. He had the vaguely Asian features of her former favorite student without the suggestion of brilliance that he had occasionally demonstrated. As for how they met, Hugh was the hygienist at her dentist's office. To Quasia's great surprise, they got along great, and it was hard to think of any of it as trouble. In fact, there was even a hint of magic there from time to time, which quite exceeded her expectations at that point. She knew quite well what she was doing to her dream a year later when she agreed to marry Hugh. As his wife, she continued to dream about her stellar career as a math professor, but no longer fantasized about her colleagues' romantic reactions to her success. (She had started feeling silly about that part of the dream anyway.) And, she was happy in her marriage, never regretting the decision ... except maybe when Hugh quit his job to become an herbalist 'prescribing' capsules of 'medicinal' extracts to gullible (in her opinion) patients.

It took longer for her to give up her expectations of the students. The worst students were hopeless beyond her comprehension, not only unable to pass the classes, but seemingly uninterested in even trying. The best students were terrifically bright, often able to go way beyond the minimum she would have expected of an A student, but even they were disappointing to her. The problem is that they viewed her not as a scholar to admire, not as a representative of what they aspire to be themselves, not even as a resource of information, but only as the creator of tests. She was nothing to them other than a part of a human obstacle course that stood between them and their degrees. Occasionally a student would show what she considered to be the proper respect, but she was now certain that this was only one of the many techniques the disingenuous students had devised to navigate the "obstacles." There was one student who had briefly raised her hopes that he might be different, one who seemed to be in so many ways like the students of her youthful fantasies, but then Quasia caught him cheating on one of her tests. Of course, according to the school's strict honor code, he was immediately expelled. She never saw another "dream" student after that, and frankly, she had stopped looking.

Okay, it is true that this was quite a disappointment, and it makes her job a lot less enjoyable than she had imagined, but she had come to grips with this problem years ago. So what else could be the cause of the problem?

Suddenly, it becomes clear, and Quasia gives herself the mental equivalent of a kick in the backside for being susceptible to something as mundane as a midlife crisis over the fact that she is not at the top of the research food chain. Moreover, now that she thinks of it, the fundamental cause of the crisis is clear as well: it is the stinging "also see" from the review article in the *Notices*.

While reading that article several months ago, she was enjoying it both as an overview of recent developments in her field and as a sort of "family reunion," bringing together—if only in print—all of the members of the representation theory family. However, when she got to the section of the article that described the area of the field in which she was directly involved, she was shocked not to see any mention of her own work ... until the summary paragraph at the end of the section where it said simply "See also references 19–33." References nineteen through thirty-three, of course, comprised her entire professional career. The author of the review might as well have written her a personal letter saying "By the way, Quasia, you're nothing but a footnote in the history of mathematics, not even worth mentioning by name" because that is how she subconsciously interpreted it.

"Quasia!" the bald physicist calls condescendingly, waving his hand just inches in front of her face. "Quasia, that was a motion, so would you call a vote please. I've got to get out of here."

"Oh, I'm sorry," Quasia says, embarrassed that she had lost track of the discussion. "I got lost in thought. Sorry."

"I guess that's what we get for electing a *mathematician* as chair," Ida jokes.

"Oh, come on now," the bald guy drones nasally, "let's not get into stereotyping."

"Yes," Pete adds with a smirk, "let's not point out that she's wearing the wrong shirt *again*."

Slowly, Quasia looks down from her colleagues, all wearing similar shirts with the logo 'Yahoo! Delivers' printed across the front in glittering letters to her own shirt that still advertises last month's sponsor.

"Yes, well, my students love it when I do this," Quasia says, hoping to sound positive. "That way one of them gets the monetary reward for reporting me to the Academic Sponsorship Office. Seeing them excited to be in class is almost worth the payroll deduction it costs."

The Monster

The set of integers includes 2, 197, −35, 0, and 10023918, but you can't list all of them. This is a set with infinitely many elements, the so-called "whole" numbers. More importantly, there is an *algebraic structure* to the integers. We know how to add them and how to multiply them. It is the additive structure that makes the integers into what mathematicians call a *group*. We know that given an integer n_1 and another integer n_2 we can add them together to get the integer $n_1 + n_2$.

It comes as a surprise to many undergraduates that numbers are not the only objects that mathematicians add. In modern algebra the integers are just one special example of a group. There are many other groups that mathematicians study, some more simple and some much more complicated than the integers. One of the simplest examples is the group that just consists of two objects, o_1 and o_0 with the addition rules $o_1 + o_0 = o_0 + o_1 = o_1$ and $o_0 + o_0 = o_1 + o_1 = o_0$. This group is important in theoretical computer science, both because of its close relationship to binary arithmetic (in which there are lots of numbers, but all written just in terms of 0's and 1's) and because of its role in representing logical arguments algebraically. However, those of a less theoretical bent can imagine it in the following very practical terms. Imagine a big toggle switch that can either be flipped to on or off, although this switch doesn't happen to be connected to any electrical circuit. The two objects o_1 and o_0 should be thought of as the two things you can do to the switch, namely o_1 means "flip the switch" and o_0 means "leave it alone". Now, think of addition of these two things as meaning "do one and then immediately do the other". For example, $o_1 + o_1$ means "throw the switch and then throw it again". Then it should be clear that the statement "$o_1 + o_1 = o_0$" is really true. Throwing the switch and then throwing it again does have the same final result as not throwing it at all; in either case the switch ends up exactly where it started. Each of the rules of addition can be interpreted in this way, and so the switch is a *model* of this group.

A more interesting group can be modeled by the following situation: imagine a flat, square piece of wood with holes drilled through each corner, and a table top with four nails in it so as to fit into the four holes in the wood. Mark the corners of the square (with numbers or colors or whatever you might want) so that you can tell them apart and place it on the nails. Now, consider the group of all of the ways you can move the piece of wood and replace it on the four nails. In essence, there are only two actions you need to consider: r that rotates the square counter-clockwise by 90 degrees, and f that flips the square over so that the bottom two corners are now at the top and vice versa. Again, addition means do one and then the other. So, for instance, $r + r$ means rotate, then rotate again ... this is a rotation of 180 degrees (another element of the group, but as you see, one that can be written in terms of r). Similarly, $r + f$ means rotate and then flip. Interestingly, however, the order now matters! Rotating and flipping ($r + f$) is not the same as doing them in the other order ($f + r$). To see this, keep track of where the corners end up. In the first case, the top right corner ends up in the bottom left, while under the second action, it ends up right back where it started. So, $r + f \neq f + r$. In fact, as you can verify if you think about it a bit, $r + f = f + r + r + r$! The fact that such equalities exist means that there are not as many different elements to this group as you might think ... this group has only eight elements all together. No matter what you do, any complicated sum of r's and f's will turn out to be one of eight different possibilities.

Okay, so mathematicians have been studying groups for a hundred years. Like a field of zoology, algebra involves both studying and classifying the beasts that are algebraic groups. One obvious classification is to distinguish between infinite groups—like the integers—and finite ones like those above. Among the finite groups, every possible group that could exist has been classified. Nearly all of the groups fit into a nice category, but then there is a category called the *sporadic groups*, which is just a way of saying "all of the groups that don't fit into any of the other categories." The sporadic groups are the duck billed platypuses of algebra, neither this nor that. It is remarkable, but algebraists really know all of the sporadic groups. Most of them are small, having less than 100 elements ... but then there is a single *huge* one, a group so large that the other sporadic groups seem tiny in comparison.

In fact, it has exactly
$$2^{46} \cdot 3^{20} \cdot 5^9 \cdot 7^6 \cdot 11^2 \cdot 13^3 \cdot 17 \cdot 19 \cdot 23 \cdot 29 \cdot 31 \cdot 41 \cdot 47 \cdot 59 \cdot 71$$
elements! What would *you* call a 10 billion pound platypus? Exactly! And so mathematicians have named this group *the monster*.

Many interesting discoveries have been made about the monster, including deep relations to algebraic geometry and number theory that go by the name of *monstrous moonshine*. But one thing still remains mysterious: the existence of the monster itself. Where does it come from? Why is it alone as the only huge sporadic group?

To be able to answer these questions is the "holy grail" of Quasia Fine's area of research, but how could one answer such philosophical questions using mathematics? There are several possibilities, but one seems particularly attractive: find a model, like the switch and the square piece of wood in the earlier examples. Find a geometric object whose symmetries are precisely the monster group, thereby producing it in a natural way. But despite the fact that many hundreds of brilliant minds have pondered this problem for decades, no such model has been discovered. The monster remains a mystery. It could be though of, in rather loose terms, as being like the Loch Ness Monster, who despite the faint outline that can be viewed in many photographs, has never really been seen by scientists. In the same way, the monster group is interpreted by many mathematicians as an indication that there lurks a geometric monster, a high-dimensional object, hiding beneath the surface of algebra, but no mathematician has ever really seen this monster itself.

A Verifiable Monster Sighting

The title of this week's Algebra Seminar ("Flatulent *D*-modules with Support in the Gentrified Weyl Chamber") does not really sound that interesting to Quasia, but she knows her attendance is expected. So, she picks up a few pieces of scrap paper and a pencil, hoping that she will be able to sit near the back where it will not be so obvious that she is working or doodling instead of taking notes. However, she is 'saved by the bell' when she is interrupted on her way out the door by the ring-ring of an off-campus telephone call.

"Professor Fine?" says the voice on the phone. She knows immediately that it is not one of her students. For one thing, they hardly ever say "Professor," preferring to call her "Doctor Fine," "Quasia" or (worst of all) "Mrs. Fine" as if she were a high school teacher. Moreover, there was something old about the voice on the phone, old and tired. That was not a description of any of her students this semester.

"Yes, this is me."

"I'm sorry to bother you," he continues, "but I am a great fan of your work, and now I think I have finally made a discovery of my own that I think might be of interest to you as well."

"Yes, well, if I may ask, *who* am I speaking to?"

"Oh, sorry about that." After a pause and a muffled cough, he explains "My name is Gordon Klein. I'm not a mathematician, not professionally anyway, but it has long been a hobby of mine. Over the years, I've proved a few little theorems, but I kept them to myself because I did not think they were important enough to bother er real mathematicians about, but now I've found something that I'm sure will interest you."

"Great," thinks Quasia, "a crank. Just what I need right now, a lunatic who thinks he's brilliant." But, either out of politeness or curiosity, she merely says "Oh, and what have you found?"

A pause.

"I've found an explicit, canonical and minimal geometric model of the Monster group."

That is a surprise. It's not the usual crank problem. Usually the cranks have disproved the continuum hypothesis, shown that there is no such thing as non-Euclidean geometry, verified the Goldbach conjecture ... but even understanding what it means to find the Monster group in a geometric context was a task beyond any of the cranks that she has encountered. Could this be for real? She decides that it can not, and her skepticism is apparent in the way she says "I'm sorry, but I really have to run and catch a talk. Perhaps you could ... "

"I know, I know, you think I'm a crank. I have foreseen this potential problem, but I can easily convince you that I know what I'm talking about."

"Okay," she says, fighting to remain skeptical, though the man on the phone sounds so very believable. "Shoot."

"One of the things I have found as a *consequence* is an unexpected formula relating the coefficients in the *PPQ*-series for the monster."

"Oh. And what is it?"

"Take the coefficients of the ith and jth terms and the coefficient of the $i^2 + j^2$ term ... they satisfy the Ogden-Nash-Hamilton formula."

"They satisfy *what?!?*" It is not that she did not hear him, or that she doesn't know this formula well. It is only that this sort of relationship is so unexpected ... but he did *say* that it would be, didn't he? The Ogden-Nash-Hamilton formula of stochastic combinatorics ... .why would it show up here?

"Yes," he says, sounding like a sympathetic father comforting a small child, "I know, it is hard to imagine. But, try it out and you'll see that I'm right."

Quasia pushes some papers off of her desk to make room for the computer keyboard and begins typing. First she loads the package `MonstMoonshine.sap` into the symbolic algebra program, and then also `StochCombinator.sap`. She has never loaded these two packages at once, since there was no reason to expect them to have anything to do with each other. But, having put them both there at the same time, in essence, having *taught* the computer everything it needs to know to be able to work with both theories, it should be easy enough to write a short program to check his claim.

After two-and-a-half minutes of phone silence while she types in her program, Quasia says "Okay, I'm ready to run it now."

She presses return and waits. She has programmed the machine to pick a random positive integer i and another integer j and check whether the wild claim of the mysterious phonecaller is true. Amazingly, it *is* true for the first pair of numbers picked, but of course that could be a coincidence. But now, the computer also informs her that he was right about the second pair of numbers, and the third, and the fourth, and the fifth, ...

This is not a mathematical proof, but it is damn convincing. This guy on the phone certainly knows *something* about the Monster that Quasia would never have guessed.

"My god, you're right. What did you say your name was again?"

"I'm Gordon, Gordon Klein."

"And why are you calling me?"

"As I've said, I have read your papers and found them all to be beautiful and inspiring ... you're one of my mathematical heroes. Oh, but wait, maybe that isn't what you meant. You mean *why* am

I calling? It is because I want your help in preparing this result for publication. I've taught myself math, but I have no idea how you write or submit a real paper!"

"Well, I am really flattered, Mr. Klein, as well as being really intrigued. Could you maybe e-mail me a copy of the result as best as you can write it up? I'll take a look and ... "

"Ho, ho, I'm sorry, Professor Fine, but I'm an old man. I don't know a thing about computers or e-mail. I'm also a bit too old to be able to come down and visit you at your office, even though it's only a short drive away. No, I was hoping that I'd got enough of your interest that I could convince you to drive out to New Jersey to meet with me at my home. Can I give you the directions?"

In ten minutes, she is in her car heading over the George Washington Bridge.

Late Again

Something is definitely wrong with the directions that Quasia is trying to follow. Westford Road ends in a dead end, not a three-way intersection as the directions say, so she can't find Holbrook Lane where Gordon Klein's house is supposed to be. After driving back and forth for ten minutes looking for the intersection of Westford and Holbrook, she finally asks for help from a woman jogging by. Jogging in place, the woman informs her—as she already knows—that there is no such intersection. However, if she goes back to Chevy Chase Street and turns right, followed by an immediate left on Houston Court, she will eventually hit Holbrook Lane. These new directions work better than the first, and she soon finds herself driving slowly down Holbrook looking for number 89. Since she's now driving past number 55 and the numbers are going up, she figures that number 89 should be somewhere in the next block; somewhere up ahead, near where that ambulance with the flashing lights is parked.

As it turns out, she was right. The ambulance is parked right in front of number 89. As the ambulance crew close the back doors and drive away, Quasia walks up to the man who remains standing on the curb.

He looks to be about thirty-five years old, wearing khaki pants, sandles and a t-shirt and when she gets close enough, Quasia realizes that the man is crying.

"I'm sorry to bother you," Quasia says without really meaning it, "but I was looking for Gordon Klein. Do you know where I could ..."

"I'm Mr. Klein," the man says, "and you are ... ?"

Quasia does not answer immediately; she is caught off guard since she was expecting someone much older. "I'm Quasia Fine," she explains, "we just spoke on the phone."

"Did we? I'm sorry, I don't recognize the name. Could you remind me what we spoke about?"

Quasia had half expected to find out that Gordon Klein was a mathematical crank, but this is *too* much. He must be some kind of complete *psycho*! "Why, you said on the phone that I was one of your heroes and practically begged me to come out here! I can't believe that you ... wait, is there perhaps *another* Gordon Klein around here?"

"Ummm, there was," he says tilting his head in the direction of the rapidly vanishing ambulance. "Perhaps it was my father that you spoke to?"

Quasia smiles at first, because that makes more sense, and then frowns when she realizes the implications. "Your father, is he okay?!?"

"No," Gordon Jr. says, choking back a sob, "he's dead. They even hooked him up to that new fractal defibrillator thing from JAMcorp, but ... nothing."

"Oh my, that's too bad. That's really too bad. We had some important things to discuss ... about your father's math research."

"Math *research*?" he says, wiping away a few tears and brightening up a bit. "Dad and his math! Look, if he told you he was doing math research, it was probably quite an exaggeration, to say the least. I mean, the guy never even finished college. I really doubt that he even understood all those math books he was reading these past few years. Any kind of 'math research' would have been far beyond him I'm sure. I hate to speak ill of the dead, but he really wasn't *there* towards the end, if you know what I mean."

Quasia wonders, could this be an accurate description of the man she spoke to on the phone? No. No way. He clearly understood what he was talking about, and there was no place that he *could* have read about the role of the ONH equation in the representation theory of the monster. If it had been published anywhere, Quasia would have seen it herself! He must have been doing real research of some sort. It was the only explanation.

"You say that he flunked out of college, or dropped out?"

"Not exactly," he says, apparently completely recovered from the shock of his father's death. "He was drafted and sent to the war in the Pacific before he had a chance to finish his degree, and when he got back he just went straight to work ... never went back to school."

"What was it he did for a living?"

Suddenly, his eyes open wide and he says "Hey, what did you say your name was? Wait here." And he runs across the lawn and into the front door of the house, leaving Quasia standing alone on the curb. He returns a few minutes later holding a manila envelope that Quasia optimistically hopes will turn out to contain a complete description of Klein's mathematical discoveries.

"Are you a professor, by any chance?" he asks. Quasia nods impatiently and he holds out the folder so that she can see the letters "Prof. Fine" written on the front in ink. "Then I guess this must be for you. Dad died at his desk and this ... I suppose this must have been the last thing he did. He was holding it in his hand when we found him. Here, it's yours."

Talking to Manhattan

Although her original plan was to drive back to her office and read the contents of the envelope there, Quasia finds that she simply does not have the patience for that. So, she pulls over at a rest stop along the Pallisades Parkway and takes a seat on the bench that the state of New Jersey has conveniently provided. Contrary to the expectations of those who arranged for that bench to be placed there, Quasia sits with her back to the lovely view across the Hudson River, choosing instead to face an empty, gravel parking lot.

She rips open the top of the envelope and gently pulls out the stack of handwritten papers that it had contained. On top is a brief note that says:

Professor Fine,

I know that we discussed this when you visited me, but I am putting this here as an important reminder. Please remember *not* to tell anyone else about my ideas. Perhaps you will think me either too shy or too paranoid, but it has taken me several months to get myself to approach you about this, and I would never have done even that if I thought that anyone other than you would see my work before I am ready to release it. I have told nobody else about these ideas, and I beg you to keep your promise to me not to mention me or my work to anyone else either ... at least not until I feel more confident.

Thanks again, Gordon.

Quasia does not even read the note; she simply lets her eye pass over it quickly and, noticing that there is no mathematics discussed, she moves on to the next page. The second page is rather light, merely summarizing all of the usual notation and the usual definitions. But, on the third page, it very quickly gets to the heart of the matter, describing a particular geometric object M (as the inverse limit of the fiber bundles over the Segre embeddings of ...) which the notes claim has exactly the monster group as its discrete symmetry group.

The proof that the symmetry group of M is the monster is rather long, and Quasia does not read it line for line while sitting there alongside the highway, but she gets enough of an impression of it to be pretty certain that it is right. She looks back over her shoulder and says to the distant island of Manhattan "I think its right."

Doing the Right Thing

It is right. The eighteen pages of beautiful mathematics that Quasia is rifling through with her thumb and forefinger contain an explicit description of the geometric object that she has dreamed of finding for years. She's read through it carefully two times in her office, and once not so carefully on the bench by the Pallisades Parkway, and she has only found three things wrong with it. Firstly, there is a "typographical error" (or whatever you would call the equivalent in the case of a hand written document) where Klein has written Q when he clearly meant q. Second, and more importantly, the second lemma fails to mention the fact that π_1 has no torsion, but Quasia is easily able to fill in this hole and verify that it does not nullify the main conclusion. Finally, the main thing wrong with it is that she had absolutely nothing to do with it. It does not build directly on any of her work. The author did all of the work without asking her assistance. It is the discovery of the late Gordon Klein, an old man without even an undergraduate degree whom Quasia has never met.

So far, however, Quasia has not consciously had any illicit thoughts about appropriating the idea as her own. She has read it merely to make sure that it is correct, and it is, and her intention is to get permission from Klein's survivors to submit it for publication. Unfortunately, that turns out not to be as easy as it sounds. There is no listing for "Gordon Klein" at that address in any of the on-line phonebooks or in the paper versions at the library. There are a good many Gordon Kleins at other addresses in the New York area, but none of them that she has been able to reach by phone seem to be the man that handed her the glorious manila envelope a few days ago. As a last resort, she drives back over to Jersey, but Klein's house is dark and empty, with a "For Sale" sign out front. She tries talking to the neighbors and calling the realtor, but the neighbors say they didn't know the residents of the house and the realtor says she is not at liberty to discuss the former owners.

It is clearly not going to be easy to find out anything about Gordon Klein, but Quasia does not feel right about sitting on this result. So, she writes a brief note explaining the unusual circumstances under which she came to be in possession of the brilliant manuscript and puts it, together with a photocopy of the paper, into a large manila envelope addressed to the editor of her favorite journal, the *Journal of Advanced Mathematics*. Very shortly, she believes, it will be in the hands of some stunned editor and Gordon Klein will have made his place in mathematical history. All she has to do is walk over to the math department, drop it in the 'outgoing mail' box on the secretary's desk, and it will be out of her hands.

Coffee Break

Of course, it is a proven mathematical fact that it is impossible to go from one building to another in Manhattan without passing a coffee bar. Quasia actually passes two on her way to the office. At least, she *would* have passed two if she had actually gone to the office ... but things don't always work out as planned. Instead, she decides to stop in at "Jumpin' Joe's Java Joint." The line is surprisingly short for this time of day, there are only two people in front of her. So, she soon orders her non-fat mega-mochalatte and moves over to the barista's counter to wait for her sugary caffeine fix.

Waiting ahead of her in line is a man with a particularly distinctive body. He is rather tall, though only from the waist up. That is, his legs seem to be the legs of someone of less than average height, but his torso and head are so long that they more than make up for it. Moreover, he is also only *fat* from the waist up. The combination of the tremendous "love handles" hanging over his belt and the short, thin legs are a very unusual combination. So unusual, in fact, that even though Quasia has not seen the man's face, and the only person she *knows* with that body type is supposed to be in Brazil, she is certain that this must be her former classmate, Jack Polynomial.

Oh god, Quasia thinks to herself, *don't let it be Jack Polynomial.*

The oddly shaped man is chatting away with the pretty young barista with the stud in her lip as she prepares his beverage. If he were flirting with her, she would know that it is *not* Jack in front of her. But, he is not behaving in a flirtatious way. Rather, his manner implies that he has something to say that is so wonderful that he cannot help but share it with everyone around him.

Then he says loudly, "Why, do you realize that each of these coffee machines here is now equipped with a microprocessor based on nonlinear dynamics?" He turns to the woman who was served before him, who is now adding some vanilla flavored powder to her coffee of choice, and continues. "I can't say for sure, since JAMcorp keeps the details of their products secret, but I bet it is based on the ideas of my PhD thesis from '97!" The woman drops the vanilla powder, spilling some on the counter, hastily puts a plastic lid on her cup, and rushes out of the storefront as quickly as she can. The girl behind the counter has also disappeared, momentarily, because she has found that the soy milk dispenser at her station is empty.

Jack turns towards Quasia, his only remaining audience, and continues nonplussed. "Who would have thought that math research in such an obscure ... Quasia?!? Hey, I thought you were here! Oh, it's great to see you."

Quasia stands motionless as Jack gives her a hug and says "I am here, Jack. You thought I was here, and I am. What a wonderful coincidence." Since he is still hugging her, he does not see the way she rolls her eyes, but the tone of her voice is unmistakable.

So, he lets her go and gives her a lopsided smile that looks like a wink, and he says "I mean, I knew you were at a school in New York, and I *thought* it was this one, but I wasn't sure. Anyway, I'm here for the Gangsta Math Slam ... "

Quasia frowns, "Gangsta Math? Jack ... "

"I know," he continues, "it's what you've got to do in these days of the PFR. If it doesn't get the public's interest, you can't get funding to work on it! Anyway, I'm still doing my own research on the side as well ... as a hobby. I'm meeting with Victor Fields—but, you must know *Vic*, he's your department head, right?—and I was hoping I would run into you. I'm so glad I did."

"Why?" Quasia says, surprising even herself. She even thinks of apologizing for her rudeness, but is interrupted by a light tap on her shoulder.

It is the barista with both of their drinks. "Why don't you two lovebirds continue your conversation over there? Or maybe even *outside*, okay?" the girl says.

They take her sarcastically offered advice and sit at a table in front of the store. A tremendous green and white striped umbrella protects them from the sun, and a cheap plastic fence—molded into the shape of pickets with an unreasonably visible wood grain pattern—protects them from the passersby on the street.

"So, how are things at IMPA these days?" Quasia asks politely.

"Quasia," Jack says with his characteristic sincerity, "I know we were never the best of friends, but I don't remember ever doing anything to justify ... I mean, if I have ever done anything to hurt you, I'm sorry."

"No," Quasia says, "*I'm* sorry. I really shouldn't have ... it's just that ... "

During a prolonged silence, they each take a sip of their drinks, and Quasia thinks to herself: *It's just that I hate you because you are so damn good at math, so amazingly brilliant, and I'm a walking footnote.*

Jack breaks the silence by saying "You know, Quasia, I really like your work."

Did he hear what I was thinking? Did I say it aloud?

"Thanks, that's really kind of you." She smiles for a moment, but then realizes that there must be some bad news coming. *You idiot, he's not complimenting you, he's buttering you up. Watch out!*

"I really mean it, your papers on the monster were really cute, and they got me to thinking about it ... "

Cute? My papers are CUTE?!?

"... in fact, I've got some results of my own that I think you'll be interested in."

"Results on *the monster?*" Quasia says, wide-eyed with fear.

"Yes," Jack continues optimistically, "on the monster, but not really directly related to anything you've done. Still, I was hoping we could talk about it while I'm here and maybe we could ... "

Unintentionally, or at worst with subconscious intent, Quasia now spills her coffee. Most of it lands on her dress, but some of it drips off of the table onto Jack's jeans as well. "Shit," they both say reflexively at the same moment. Then, Quasia jumps back from the table and runs away in the direction of her office.

"'Cute' is *nice*, Quasia!" Jack calls after her, "I really mean it, your work is good!" Then, after she has gone into her building, he uncharacteristically mutters to himself "Oversensitive bitch!" with a shake of his head. "She'll never change."

A Flashback

Coincidentally, it was exactly seven years ago to the minute that Quasia Fine last spilled a cup of coffee on herself sitting at that same spot. (Well, to be honest, it was perhaps a few inches to the right since the former owners used larger chairs, and it might have been a minute sooner or later ... but you must admit, it sounds better to be definitive.) That was back when she was a *happy* math professor. She had just received word that her NSF grant proposal had been approved (it was also back before the Public Funding of Research Act required scientists to advertise for donations from average citizens in order to fund their work). And she was grading the make-up test of her favorite student, James Moy. She was quite happy indeed.

However, she found herself disturbed by James's answer to question number 3. It seemed to be just a bit too similar to the answer that was given on the *other* student's make-up test, that she had just finished grading. At first, her inclination was to suppose that it was the other student who cheated. After all, this was *James's* test. But, that theory did not stand up to further scrutiny. For one thing, James took his test the day after the other student. Moreover, she had written two versions of the test for the very purpose of catching cheaters, and the two make-up tests had different versions of question 3.

James's test showed the other student's answer to a question that did not even appear on James's test. There was no escaping the conclusion: James had cheated!

Why Not?

"... and the song you just heard was 'Sudden Illinois' by Ned and the Welchers. Now, thanks to the brilliance of your national legislature, we have no choice but to play you five minutes of biologists begging for your money. First up on the PFR hit parade we have a PhD/MD from

Madison, Wisconsin who will tell us about ... " (Click).

The traffic on the L.I.E. is stop-and-go as she heads home. Normally, she would listen to the radio in such circumstances. But, Quasia can not listen to the car radio right now. She is too busy being confused by her own neuroses.

The envelope containing Klein's article and the note she had written to the editor remains unmailed, lying casually on the passenger seat next to her, amid a pile of other papers and books. But those other reading materials are merely camouflage, grabbed quickly off her desk before she left work a few hours early. They were taken when she was still pretending to herself that Klein's article was going to be mailed as soon as she got home and that she would spend the rest of the night reading.

By this point, however, she is openly considering a questionable alternative: submitting an article on this result with her own name listed as an author ... possibly as the only author.

She even attempts to make such an action sound justifiable with pseudo-logical arguments like:

"He had a good idea using that Segre imbedding like that, but I think I already see a better, more professional way to do this using the tri-secant identity for highest-weights in $gl(\infty)$. Yes, that will be *my* construction. There is really no need to mention his Segre approach if I do that."

"He wanted me to promise not to mention him or his ideas to anyone until he was ready. Obviously, he'll never be able to tell me he is ready. So, instead I can publish my version and thereby keep the promise. That would certainly be the way he wanted it to be."

"Well, I was his mathematical hero. It was my work that inspired him to do this."

"He was really just a crank who stumbled upon something brilliant through mere luck. My role will be to have found him and to rework it so that it is real mathematics rather than the ravings of a sad old man."

So, it could not be called "unexpected" when Quasia drives past the mailbox on the corner at which she had intended to mail the envelope and instead drives straight to her house.

She is relieved to see that Hugh's car is not in the garage and rapidly keys in her entry code (the first three terms in the *PPQ*-series for the Monster) at the front door. Rather than opening, however, the door beeps at her. She almost keys the code again, assuming that she has accidentally entered it wrong, when she notices the message appearing on the display:

"Message from SekYouritySystims:

Re: Update Required

Due to a new form of attack to which prior SYS entry systems are susceptible, access is disabled until you download Version 13.0 which includes the JAMcorp VerID patch.

Enter Root password to update now."

So, with no choice if she wants to gain access to her house ever again, Quasia keys in the Root password (which is the MR number of her first published paper) and waits for the system to be updated.

Flashback II

According to official policy, Quasia had no choice but to report evidence of a student cheating to the Honor Board that judges student violations of the school's strict honor code. Of course, there are instances in which professors fail to do so—such as if they are sympathetic to the student's situation or (more frequently) if they feel that giving the student their deserved punishment is not worth all of the paperwork that it would involve. Quasia, however, had greater motivation than usual to pursue "justice." She was still angry at James for having *appeared* to be such an ideal student, for getting her hopes up. In short, she was really seeking revenge of a very personal sort when she followed the rules and contacted the chair of the Honor Board.

At the hearing, James admitted to cheating. Of course, there were excuses for the board to consider—there always are. Several of these were quite familiar to the members of the board from other cases they had heard. James was doing well in the class before this test, apparently without cheating, and so probably would have been able to pass this test as well if it hadn't been for certain extenuating circumstances. They had heard that before and it had never meant anything to them in the past. The extenuating circumstances in this case happened to be his father's sudden demise due to an undiagnosed heart condition that prevented him from properly studying for the test. Even that had come up several times before and had never given the board reason to doubt the justice of the required punishment.

You see, although the honor board acts as if they have some sort of decision to make and asks a lot of questions of all parties concerned, since James admitted to the cheating, the rules really left them no option other than expulsion.

But, this case was troubling to many of the members of the board, both the students and the faculty. Of course, as James explained, there was tremendous pressure on him to do well on the test as it could possibly affect his entire future. That was something which did not need to be stated, it was taken for granted by the board and precedent had already determined that this was 'no excuse' for cheating. Still, the board understood well that such pressures exist, and that is why they felt that it is partly the responsibility of the professor to help the students avoid temptation by making cheating at least reasonably difficult. They should sit there while proctoring the exam and look carefully at the students ... or at least give the impression that they are looking carefully. They should keep students from sitting on each other's laps and inquire about any mumbling from the test takers in the back of the room. In short, they should do *their part*.

What made this case different than all of the others the board had considered was that Professor Fine had *accidentally* made cheating too easy. In particular, since James had missed the test—in order to visit his sick father who died before he arrived—he took a make-up test in Quasia's office. She very trustingly had let him take the test alone, sitting at her desk, while she covered a stats lecture for a sick colleague. She very negligently—in the opinion of the board members—had failed to remove from her desktop the make-up exam that another student had taken the day before. (The other student, being less-trustworthy in Quasia's opinion, had taken the test under the watchful eye of the departmental secretaries at a spare desk in the main office, but afterwards the test had been dropped on her desk and forgotten.)

Quasia, feeling that her trust had been taken advantage of, did not see this as anything other than the circumstances of James' cheating. The board, and James as well, saw it as description of Quasia's role in the violations. Unfortunately, according to the very carefully worded rules of the University, there was nothing the board could do to give Professor Fine her fair share of the punishment or to lessen James' punishment accordingly. Immediately following the hearing, James was no longer a student of the University, nor would he likely ever be a student at any reputable school again.

The day after the Honor Board hearing, she casually discussed the situation and the outcome with her friend Vic, who was then only associate chair. Vic could not understand how Quasia could sleep at night knowing that James' life had been ruined by her negligence. As he described it, the rules that led to James' expulsion were guidelines to be followed only when the situation was "normal." Once the situation was so unusual as to have not been considered by those who created the rules, improvisation had to replace the rules. He made it clear that had *he* been the professor rather than Quasia, his improvised approach to the situation would not have included any serious punishment for James.

Quasia could not quite understand why Vic "took the side" of the cheater.

Legal Division

It has been less than one month since Quasia submitted *her version* of the result to *The Journal of Advanced Mathematics* for publication. Despite all of the automation that has simplified and expedited many steps of the review process at professional math journals, the usual wait time before hearing about whether a paper has been accepted is between 6 and 10 months. So, she is not at all surprised that she has not heard anything yet.

Still, since the result is *so* wonderful, and since Vic Fields is an associate editor of the journal—even if in another field of mathematics—she wonders whether this has anything to do with the fact that she has been summoned to the office of the department head. She imagines him congratulating her on the great results, saying that it would have made her a good candidate for a Fields Medal if not for her age—quite a bit like her youthful fantasy minus the steamier aspects. This seems less than likely once she sees who is waiting for her in the office and the serious—no, *angry*—expressions that they share.

As the chair of the math department, Vic gets an impressively large office with a marvelous view of lower Manhattan out the window. He is sitting behind his large black desk, with his sneakered feet crossed and resting on top as usual. Normally, this position includes hands clasped behind his neck and a broad smile that fits perfectly between the wrinkles of skin that he had gained over the time Quasia has known him. Now, however, his arms are folded across his chest and the wrinkles seem contrived surrounding a frown. There are also two visitors in the office, sitting in a pair of chairs that block the marvelous view. One of them is Jack Polynomial, whose short legs barely touch the floor, looking as if he wants to bite her head off. The other is a man with greying sideburns and wire rim glasses, looking equally serious.

She does not know who he is, but his suit and tie are a good indication that he is not a mathematician.

"Vic, Jack, Sir ... " she says, nodding to each in turn, "is there something wrong?"

"Well, I certainly hope not," Vic says looking past her as if he is unable to look her in the eyes. "I'm thinking that you can probably just explain yourself and this whole mess will just go away. If we followed normal channels in this situation, we'd not be just getting together casually like this, but I think that a situation like this calls for ... well ..."

"Improvisation?" Quasia suggests.

"Yes, that's the word I was looking for. Thanks. Anyway, I was hoping that if we just got all of the interested parties together, we might be able to straighten things out with less trouble."

"I hope so too, I guess. What exactly *is* the trouble?" Quasia glances over at the two other men who have been conspicuously silent until now.

"Professor Fine," the stranger says, straightening up in his chair just a little, "allow me to introduce myself. I'm Will Testament from the legal division at JAM and ... "

"Do you mean JAMcorp?"

"No, I'm sorry Professor Fine, I guess I wasn't clear. No, I mean the *Journal of Advanced Mathematics*. We now have some dealings with JAMcorp since we've purchased their software for managing the refereeing process, but so has nearly everyone else and in any case we are clearly two separate entities."

"The journal has a *legal* division?"

"Yes, of course ... doesn't everyone? In any case, I'm here as a representative of the journal. Even though we have no legal interest in the matter it is as a favor to Professor Fields that ... "

"Would you get to the point already!?" Jack shouts. "Quasia, you plagiarized my result on the Monster and tried to pass it off as your own. That's the point, and it explains why you were so weird when I ran into you the other day at ... "

"Wait," Vic drawls. "Wait, wait. We're not jumping to any conclusions here. Okay, Jack? Let's start with some facts. Quasia, did you submit this paper for publication in JAM last week?"

"I can't vouch for each word of the text from this distance, but yes, that looks like the manuscript I submitted last week."

"And," Vic continues, "did you download this paper to referee two weeks earlier?"

From across the desk, Quasia really cannot see the paper very clearly.

She considers saying "no," simply on the basis that she did not download *any* papers to referee during the previous month, but instead she says "Can I see that please?"

She looks quickly through the paper that Vic hands her. It looks almost exactly like Gordon Klein's handwritten draft that she had received the week before, though it is nicely typeset in TEX. However, this paper lists Jack Polynomial as the author!

"No, I've never seen that paper before," she says, sounding less convincing than she might have liked.

"I'm sorry to contradict you, Professor Fine, but JAM has records indicating that you did download this paper the day after we asked you to review it. According to the log, you signed in with your secret password from the computer in your office and downloaded this file. To do so you would have had to have accepted the agreement that describes some rules of conduct for referees at our journal. Needless to say, submitting another paper based on the material you have read is considered inappropriate. Submitting it to the same journal without any citation of the original work is simply ..."

"Look, I did *not* sign-in and download Jack's paper from the JAM site. Obviously, if I did, I would not have sent it right back to the same journal."

"But you did, Quasia. Admit it, you've really lost your mind! You abused me both verbally and physically at the coffee shop the other day with no clear provocation. Then I get a call from Vic telling me about all of this! You are an absolute nut!"

"Wait, I don't think we need to go into all of this. Perhaps, Quasia, you had been working on this same idea before you saw Jack's paper and didn't want to get scooped ... ."

"No, Vic, that's not what happened!"

"... Okay, whatever. In any case, I think I see another solution to the problem. This is obviously not the kind of result that one can figure out over night. So, I'm figuring you must have some evidence that you were working towards this result *before* Jack even submitted his paper. Notebooks, SAP files, something clearly dated from a time earlier than this month that would give you some priority on these results?"

"Not really ... I just don't work that way."

Vic's frown deepens. This had been his one hope of resolving the conflict in a reasonable manner. Now, however, he concludes that Jack's assessment of her mental state is the most likely explanation of the unusual situation.

"Quasia, I'm afraid I'm going to have to ask you to take an immediate leave of absence from your job here. I'll find an adjunct to cover your classes, and your salary will continue ... at least for the time being."

Quasia is too shocked to say anything. What can she say, anyway? Should she explain the whole story, beginning with the phone call from Gordon Klein? At this point, that would most likely only get her in more trouble. So soon after fantasizing about the success she would receive as a consequence of this paper, it was difficult for her to accept that instead she might never be able to work at a reputable school again.

An Old Friend

Walking to her parking garage after the meeting with Vic, Jack and Mr. Testament, Quasia is—to say the least—extremely upset. This explains why she does not notice the very unusual car that drives up beside her on the street, but the other pedestrians around her do. It is a Quiver, the new luxury, Hydrogen cell sports car that costs well over $1 million when loaded with features, as this one clearly is. The other pedestrians ooh and aah as it follows her down the street. She doesn't even notice when the window rolls down and the driver starts calling to her.

"Professor Fine! Professor Fine, hey, over here! Hey, professor!"

Finally, she realizes that she is being called and so she glances into the car at the driver. Without saying a word, she looks away again and keeps walking. It takes a moment for her brain to process the information it has just received: *very expensive looking car, handsome, young Eurasian driver calling me by name ... face looks familiar. He is much older now, but I know who that is!*

"James? James Moy?"

"That's right, Professor Fine. I'm glad you remember me. Look, I need to talk to you. Could you hop in and I'll give you a lift to your car?"

"I'm sorry, James, but this really isn't a good time. I'm ... I'm feeling really *sick* right now and I'd hate to throw-up inside your nice looking car, there."

"I know what's going on, Quasia, and I can help. Trust me. Hop in."

*He knows what's going on? How can he? I don't think even I do ...
unless ...* In a flash, she recognizes the irony of her situation as compared to that suffered by James all of those years ago. It couldn't be a coincidence ... mathematicians don't believe in coincidence.

"Great, yes. Just hold the buckle by the receptacle and it locks automatically ... magnetically! That's right. Now, hold on tight while we head over to your parking spot; this car has quite a pick-up!"

"James, how did you get this car? As I recall, you were working fast food just to pay for school."

"Fast food, yeah, that seems like another lifetime now. I hated filling those damn ketchup bottles. Well, let me fill you in on a little secret. Promise not to tell anyone? You're sitting in the car of the president of JAMcorp!"

"I think I saw her on the TV once. What's her name? Alicia Baxter or something?"

"No, no. That's my sister—Baxter's just her married name—and she's the CEO of JAMcorp. *I'm* the president ... and co-founder!"

"Oh," she says, and then thinks about it a little more. "Oh! That explains a lot. So, JAM is for 'James and Alicia Moy'?"

"Sort of, or maybe it's for 'Just Advertising and Marketing'. You see, when I told my sister—who was a hotshot B-school student—about how cool math was, we realized that all it was missing was, you know, advertising and marketing! Things have been going so well, I guess you could say we were right."

They sit silently for a while as he drives her right to her car. Since she parks in a different spot each day, she wonders briefly how he knew where to find her car without asking her about it.

"How much do you know about me? To get me into this mess, you must have known quite a bit ... about a lot of things."

"Well, it is a slight exaggeration, but really only a slight one, to say that we know everything about everyone. That's not really been our goal, but it is sometimes a useful by-product of our success. I'll admit that for quite a while, I had been watching the submissions at the journals using our software for their review process, waiting for a paper like the one that your friend Jack eventually sent in."

"I could try to convince people of what really happened, regain at least a little of my pride—get my job back."

"Good luck."

"Like, if I supposedly downloaded this paper according to the records kept by your company, how come I don't have a copy?"

"But you do, its on your computer at home right now."

"Not a *printed* copy! I always print out ..."

"Yes, there's a printed copy on your kitchen table too."

"But my fingerprints ..."

"Of course, Quasia, don't be silly. Your fingerprints are all over that paper. Face it, you know that you didn't see Jack's paper before today—at least not all nicely typeset like that and everything—but nobody else will ever believe you. I've got nearly unlimited resources, and I don't think you'll be able to beat me on this one. Besides, let's not forget that even though I may have been responsible for the set up, you really *did* knowingly send in a paper that you plagiarized. What difference does it make whether you knew it was Jack's? You'd be better off just biting the bullet, as they say."

"And then what? What does the future hold for the former Professor Fine?"

"I was thinking of offering you a job at JAMcorp, in our research division."

"Why? Why would you offer me a job?"

"Oh, I've read all of your mathematical papers ... you're one of my mathematical heroes. I really mean that. And, I suppose you could say that I would not be the success I am today if not for your influence. It would be a good job. Double the salary of your professorship and we've got great benefits."

"I'm not sure if you're doing this just to humiliate me or if you have some more dramatic torture planned for me when I accept."

"Neither, I assure you. Don't you trust me? I merely wanted to make a point, and I believe I have. Now, separately from that I am making you an offer that I think is in your best interest ... our mutual benefit, in fact. What do you say?"

"Do I really have a choice?"

"Yes, of course. You always do."

Finale

[*Announcer:*] *Welcome to another episode of "Research Rumble", the PFR game show where you, the studio audience, get to decide which of our contestants will get the big bucks to continue their studies in the topic of the day. Today, we ask the question "Why was it that animals on Earth used to be so big, but no longer are?" I mean, when was the last time you saw a dragonfly this big in your backyard, eh? Our contestants include Sid Dementiary, a geologist who would like to argue that the Earth's atmosphere lost a lot of its oxygen when its magnetic field reversed its polarity; Jean Ohm, an evolutionary biologist who will tell us why she thinks that it was the introduction of the human brain into the equations that turned size from an advantage into a liability; and finally in our quack corner, air-conditioner repairman Justin Seine with his own idea about gravity increasing slowly with time. We're going to pause for a break from our sponsor, but remember that our contestants are right now donning their protective headgear for a round of Slimepit of Despair ... so don't go away!*

[*cheerful jingle:*] *From your morning coffee to your evening bath, we're improving it all with the power of MATH!*

"Hi, I'm Dr. Q. Fine, senior vice-president for research at JAM-Corp, and I'm thrilled to be here to give you a heads up on a new product that we'll be selling in a few months. I mean, I wouldn't want you to go out and buy a laptop or PDA now, just to be disappointed that you didn't wait for the new line of small and unbelievably fast computers powered by our Monstrous Moonshine chip, the MMRT5000. You haven't seen computer power until you've seen what sporadic representation theory can do to ... (click)."

10

The Corollary

Your winters are awfully cold since you've taken a job at this big Midwestern university. Despite the advice of the experts, you have not been able to wear enough layers to keep yourself feeling warm.

But tonight, the cold does not bother you. Why? Well, it isn't the weather. It isn't the way you are dressed. It isn't even the glowing fireplace beside you. No, you know what it is.

Tonight is "the night"! The chill that normally haunts your bones has been supplanted by the hope—the *knowledge*—that tonight you will finally prove your worth. After nearly fourteen years, you are finally about to resolve the greatest open problem in mathematics.

Won't your colleagues at work be surprised? They have no idea that you were even working on it. Now, instead of you being jealous all of the time, it is their turn to be jealous of you. The ones who went to grad school with you and wound up at the Ivys, the postdoc you roomed with at the conference in Greece who publishes in *Annals* every other month, everyone who refused to date you because you were too nerdy ... they'll all hear about your triumph and be jealous of *you* for a change.

Even more importantly, you will be proving your worth to *yourself*. Admit it, as cocky as you are, you even doubted yourself sometimes. Your colleagues may not appreciate you for your brilliance, but you know they are impressed by your publication record. In this "publish or perish" world, quantity is often more important than quality. And that is a skill you seem to have mastered. Don't they realize how easy it is to double—or even triple—their publication rates? You know all the tricks in the book. Using intentionally obscure notation, you can publish the same result several times without anyone seeming to notice. Most importantly, you know the trick of choosing who to send the paper to—and when—so that you can get almost anything published. Tiny advances on previous results can look especially impressive to editors and reviewers who didn't understand the original results in the first place. Small omissions, like entire lemmas without which the paper is worthless, are overlooked by editors who consider you a friend and are busy with other things like personal problems or impending retirements. Do they even bother sending your papers to referees? You have often had your doubts.

Of course, you feel bad about all of this. But, remember, you *had no choice*. Especially since you are such a lousy teacher, keeping your publication rate up was really the only way you could keep your job so that you could keep working on the big problem. At times, you doubted yourself, fearing that you were a fraud. Fortunately, you can see now that it was not in vain.

There! "QED." Over fifty handwritten pages, it is not only your longest work, it is your masterpiece. No tricks, nothing missing. It is a work you can be proud of. And you are, you feel amazingly ... different!

Look at the flames dancing. Beautiful. Not just the fireplace, your whole living room. Did you ever appreciate the intricate woodwork before? Incredible! So many years of self-hatred wiped away in an instant. You can finally look at the world around you without wondering how poorly it reflects on you.

And, when you finally publish it ... then ... then ... then the world will look at *you*. You will be the one who makes the other mathematicians cringe, though you can't quite understand why that was once a motivating desire of yours. You will be the one whose name is mentioned among the great gods of mathematics.

But wait, don't you realize? When you are so famous, people will want to read your *earlier* works. Your friends who only browsed your papers before will want to read them carefully. Professors will require their students to read and summarize them. The other gods, who never noticed you before, will look them up to see where their new colleague has come from. Some of them surely will notice that the papers are crap. It cannot go unnoticed that a few of your papers are unworthy. Not only unworthy of a genius like yourself, but unworthy of publication at all!

Then what will people think of you?

That's right, we knew it was too good to last. Did you see what happened? It was barely two minutes before you returned to your self-hatred. That aching behind your ears, that slouch, the way you grind your teeth, they all came back as soon as we realized this necessary corollary of your beautiful theorem.

Remember the way it felt, though? Remember the flame and the woodwork? You want them back, right? Well, there is only one way. Throw your paper in the flames. Yes, and the notes too. All of them must be burnt. There must be no evidence that you were ever even working on this.

You see!? Whoosh! The fire is gorgeous and *huge*. The papers break apart and curl up into hundreds of tiny fluorescent spirals. Doesn't it remind you of that documentary you saw last year? Remember how you learned for the first time that our galaxy, a fluorescent spiral itself, will be destroyed in 5 billion years when it collides with another galaxy and how that made you feel? *Rushed!* It made you want to work on the problem. And now? Isn't it amazing! Two *galaxies* colliding; five *billion* years. And it doesn't bother you at all. In fact, it is a comforting thought. You *are* comfortable.

11

Maxwell's Equations

In the study of James Clerk Maxwell, Glenair, Scotland, 1865: I should not be sitting here writing mathematical formulas! I am not a professor, not anymore. Moreover, there are guests waiting for me downstairs. But I cannot pull myself away from this work. Oh, let them begin the meal without me. What company would I be with my mind entirely engaged in this puzzle?

There are equations here to be discovered and explored, if only one takes the time to look. We have known for a long time that electrically charged particles and magnets interact. You can move charged particles with a magnet, making electricity. You can make a magnet with an electric current. All I am doing here is writing down a formula, using the calculus as Newton would have done, to show us the mathematical expression of this interaction.

Oh, I can hear the objections already. "Maxwell has retired, he is now nothing more than an external examiner in mathematics, and still he thinks he can discover facts about the world without the use of any scientific instruments whatever! He has claimed before that he knows, without making any observations, the composition of the rings of Saturn! And now, he will tell us something new of electricity and magnetism from the confines of his study!"

Perhaps they are right. Should someone build a vessel capable of taking us to Saturn and find that the rings were solid, how could I object? "No," I would say, "you must be wrong since I have proved it to be so by logic."

But, what about these symbols I have written here. On the surface, they say nothing new. These are merely the equations governing the laws we already know; they are the knowledge of our physicists written in the code of mathematics. Yet, by manipulating the symbols correctly, could I not learn something hitherto unknown ?

For instance, could we not view these two terms as a collective whole, balanced out by this term which I could easily add here producing ... nothing new. This is just an unusual way to return to the exact same form. Well, we could also, merely for reasons of symmetry, add such a term as this here—again balanced appropriately so as to preserve the validity of the equality—and then we at least have something that *looks* significantly different than what we began with.

But wait, what is this familiar term that I see hiding within this new form of the equation? This old friend has nothing to do with magnets, nothing to do with electrically charged particles, but is instead D'Alembert's description of the violin string. How odd that this simplest mathematical description of a *wave* should appear here in my work. The vibrations of a taut string, leading to vibrations of the air, carried in a wave of sound to our ear. The beauty of the mathematical description of this phenomenon is equalled only by the beauty of the sound of the violin itself.

What can this mean? My computations would be the same were the experiment to be done in a vacuum, so it surely does not indicate sound. Perhaps it is just a coincidence that this wave equation appears in my magnetic formulas. Yes, it is definitely just a coincidence and I should leave this mess of notation to join my friends ... unless, perhaps there is a wave of a different nature that must be considered here. Is this possible? Certainly, light itself is a wave that may travel through a vacuum. Furthermore, it is now known to science that the interaction of electricity and magnetism is propagated at the speed of light.

But, this formula could not be describing the waves of *light*!?! Surely, the experiment has been done many times, with no mention of light being produced. No, I really am being ridiculous to think that the mere appearance of an equation describing a wave in one situation must be of significance when it appears in another. Besides, I am getting hungry ... ah, wait ... *frequency*!

Just as the period of D'Alembert's solution to the wave equation determines the pitch of the sound, so too should the frequency here have a physical consequence. Perhaps there are light waves produced here, but of a frequency that we cannot see, just as I seem to be unable to hear high pitched sounds that are perfectly audible to my friends.

Do even *I* believe what I am saying? Yes, I am quite certain ... as certain as I am that the rings of Saturn consist of a multitude of fragments ... I am certain that there are waves of electricity and magnetism, that some travel through space invisibly and that some are what we call light. We can even make these waves, as an orchestra makes waves of sound in the air, by controlling an electrical current! And, I am certain that I can completely describe these waves with my mathematical symbols ... but it will have to wait until tomorrow because my hunger cannot wait any longer.

12

Another New Math

"You may go in now, professor" the secretary said, indicating the open door behind her desk.

"Please wait here for just a little while," Professor Krell whispered to his young daughter who had been waiting with him. "You can keep doing your 'work'." He handed her a pad of drawing paper and a small box of crayons.

"Okay, dad!" she said smiling, and she watched him walk into the conference room. When the door closed behind him, she picked up a broken, pinkish crayon and began coloring.

Inside, the professor took a seat at one of the available chairs around the conference table and greeted the other people who were there.

"Good morning, professor," began a woman in a red sport suit, tapping her pen slowly on the table top. "You know, we have agreed to meet with you out of respect for your status as one of our country's greatest scientists, but I must tell you that we have read your proposal and find it absolutely ridiculous."

"Elizabeth! Please be polite," interrupted an older woman from the other side of the table. She turned to the professor and smiled. "All she means to say is that we here at the board of education receive many suggestions for how we should run our schools. We are very interested in hearing more about your proposal, but this does not mean that we are likely to adopt it. In some ways your proposal reminds me of the 'new math' craze that was so popular when I was younger. But, those ideas are no longer in style and your proposal does seem rather ... ambitious."

"From what we've read," said a bald man with a loud tie, "you are proposing that we begin teaching college, or even *graduate* level mathematics to our elementary school children. Is that right?"

"Well, yes and no. There are many parts of mathematics that students currently do not learn until college, or graduate school if they get that far. But, they are mathematical ideas, just like the ideas we teach to children. I don't see why one is necessarily more elementary than another."

"For instance," the woman who spoke first said, pointing her pen at a paragraph in his proposal, "you say that you would teach students 'group theory' instead of arithmetic. Now, I'm probably the only person here, besides yourself, who knows what that means—I have a degree in *math* education. Group theory is where you have things like numbers, but the rules are different. Like, maybe there is no law of commutativity."

"Could you remind me of what that means?" said the man with the tie.

"Sure," she said, leaning across the table to write a few formulas on the manila folder in front of him. "The law of commutativity is what tells us that the order doesn't matter when you add or multiply. Three plus two is the same as two plus three ... five times seven is the same as seven times five."

"Oh," he said wide-eyed, turning to the professor, "and you want us to teach the children that five times seven is *different* than seven times five?"

"No," the professor said, wondering whether he was ever going to be able to get his point across, "of course five times seven is the same as seven times five. However, among all possible groups, this commutativity is an unusual phenomenon. When you start looking for group structures in the real world, you can't just assume that they will be commutative. I think we're giving kids the wrong idea by focusing on a bizarre commutative case for so much of their schooling. It is as if we teach them only about giraffes and then never mention any other animals in their schooling. How could we fault them then for incorrectly concluding that all animals have long necks and eat leaves?"

"Okay," said the older woman, still polite but clearly skeptical. "What about this part here, where you talk about 'linear' and 'nonlinear,' what does that mean?"

"Suppose you have an equation," he said and waited for her to nod before he continued. "If you can always scale or add two solutions of the equation together to get a new solution, then it is linear, if not then it is nonlinear." The man with the tie and the older woman both rolled their eyes involuntarily and turned towards the woman in the red sport suit, hoping she could explain.

"Don't look at me! All I know is that the classes I took at school were always about linear things. All we learned about nonlinear was that it was too hard for us to think about!"

"That's exactly my point!" said the professor, pleased by the fact that he was going to be able to make his point even if nobody understood him. "We hide nonlinearity from people, even from students in college who are supposed to be learning math, as if it was impossible to understand. But we deal with nonlinearity everyday in the world. Reality isn't linear. Look, I can be sitting in this chair or you could be sitting in this chair, but we couldn't *both* be sitting in this chair. That's nonlinearity. Of course, quantum physics wants us to think that things are linear, it is all constructed as a linear theory, but then *poof* you need some sort of magic to explain what happens. You can tell that we're making a mistake when we can't tell the difference between our own scientific theories and appeals to magic."

Nobody was listening to him anymore. At this point he was just ranting, and he knew it, but he couldn't stop himself.

"And *why* is quantum physics such an incomprehensible mess? It isn't because reality is a mess, reality is clearly very nice. It is because of the way we learn. We learn about commutativity and linear systems so much, that when we finally get up to the point at which we are supposed to learn about how things *really* work we are not able to build any intuition!"

"Professor Krell! I'm afraid we don't have any more time for this. Do you have *any* idea of how a child's mind works? Of how a child learns? Do you have any idea of how a child would react if you tried to teach them all of this stuff instead of the math they will really need in life?"

"Actually," the professor said in a strangely cheerful voice, "I do."

"Daaaddy," called a voice from behind the door. "Is it time yet?"

"Just a minute, honey." he called back to his daughter.

The three educators laughed and the man said "I think that your daughter is getting impatient. She is ready to go home now."

"To go *home?*" said the professor. "Oh no, you misunderstand. She is impatient to come in here to meet all of you. You see, I've already begun teaching her mathematics in the manner that I've proposed to you."

"And how *old* is your daughter, professor?"

"She just turned five last week. I figured we had a choice: she could either fill up that little brain of hers with the names of all of the Pokemon monsters or she could learn something important. So, instead of teaching her the natural numbers—1, 2, 3, and so on—as most parents do, I gave her something a bit more challenging ..."

"And you'd like us to judge the results for ourselves," said the woman in the red suit. "Sure, I'm curious."

"Daaaddy," called the voice from behind the door again. "Is it time now?"

"Yes, you can come in now," he called back.

The older woman got up and started towards the door to open it, but before she could get there everyone was startled by a flash of light and an odd pinging sound. A small girl appeared—seemingly from nowhere—in the chair next to Professor Krell who just smiled and nodded to himself.

13

The Center of the Universe

"Making any progress, Allen?"

"A bit."

"Which kind of 'bit,' Shannon's or Tukey's?"

"Neither, you unbelievable math geek! I just meant 'a little bit,' which in this case was actually a euphemism for 'not at all'."

This was not an unusual conversation for room B32 in the Mathematics Department at this time of day, but Conor detected a painful hopelessness in Allen's reply. It was a feeling he remembered well, with almost masochistic nostalgia, from his own first year as a graduate student.

"I know you think you hate it now," Conor assured his office-mate, "but trust me, next year you'll really appreciate the hell she's putting you through."

Allen tried, but found he could not imagine ever appreciating the sleepless nights and emotional anguish caused by the take-home exams in his analysis class. Any attempt to answer one of the several questions on each bi-weekly test began with a state of confusion in which you doubted that you understood what was being asked followed by a long period of understanding the question but doubting that you should have ever chosen to get a degree in mathematics in the first place. In the end though, he had to admit that after each previous test there had been a sense of accomplishment and a very deep understanding of the subject. Even so, that didn't change the fact that at this very moment he had no idea how to answer two out of the three questions and little confidence that his answer to the third was even close to being correct. In response to Conor's comment, he just nodded and grunted.

Conor returned to his self-imposed task of the evening which was to compute digits very far out in the expansion of the number π in hexadecimal. He also had homework to do, for his differential topology class, but for now he was so absorbed in the useless computation he had undertaken that he had no mental energy left to think about it.

"Can I ask you something?"

Conor looked at Allen over the computer monitor and smirked. They both knew that Allen was not allowed to get any assistance on the exam. It was a matter of honor. "Sure, ask away!"

"Does Pearson *always* teach real analysis?"

"She taught it last year when I took it, and the year before that when my former officemate had it. Before that I don't know, but I definitely get the feeling that it is *her* class and she really knows how to teach it."

"If you loved it so much, Conor, how come you didn't wind up in analysis? Your advisor works in number theory, so you won't be seeing much analysis there."

"There's more than you would think. Besides, I just love some of the results in number theory. Like this thing I'm working on now ...."

"You're doing real research already? You're just a second year grad student."

"Nah, it's not research. At least, it's not *my* research. Some famous guys already figured this out ... I'm just playing with it. Listen to this! You can calculate any digit of π that you want, like the two-hundred millionth one, without computing any of the previous digits. So, what I'm doing now is ..."

"Conor, come on. I'm supposed to be doing this take-home right now. Could we talk about this tomorrow?"

"Sure, sorry. I forgot."

Allen looked back down at the photocopied question sheet, trying to understand exactly what a Riemann surface was and how it could possibly be used to prove what he needed to prove in problem number two. Professor Pearson had described the surface as being somehow like a parking garage, whatever that meant, and the definition in the textbook was too abstract for him to think about. He was sure that his classmates, with whom he was not allowed to discuss the test, were doing very well in answering this question. He had looked over his notes, read the appropriate sections in the textbook and in another analysis book he was using as a second source; he had tried everything imaginable to get himself to understand what he was supposed to do, and he knew that he would throw up if he had to try again.

"What was that you were saying about π, Conor?"

"You sure you want to talk about it now?"

"Yeah, I'm not getting anywhere on this."

"It always seems that way, Allen. And it always seems like you're the only one who isn't getting it too. What used to happen to me was that I'd give up, go to bed, and then wake up two hours later with an idea that turned into another idea and eventually became a correct answer right before the test was due."

"Sounds like a good plan to me. Okay, I give up. Now I'll give it a couple of hours and see if I get anywhere. So, what *were* you saying about π?"

"This is really cool. Check it out. You know that the decimal expansion of the number π goes on forever without getting into a repeating pattern, right?"

"Sure, I may not know what a Riemann surface is, but I do know a lot about the number π. I used to read about it for fun when I was an undergrad. That's one of the things that got me to get a degree in math in the first place! Like you said, it's *irrational* so its decimal expansion isn't the kind that just repeats itself ... "

"Right. That's why it is so amazing that this algorithm can compute the nth digit in the expansion. *Just* the nth digit, for any n, without generating any information about the earlier digits."

"Are you pulling my leg?" Allen inquired sincerely. "I've never heard of such a thing and I can't imagine how it could work."

"Look, I don't fully understand it either, but it's pretty new so you wouldn't have read about it when you were an undergrad, and it's not a joke. Here check it out." Conor tossed his friend a copy of the paper *Algorithmic Optimization of a Method of Borwein et al.* Allen looks at a few pages and finds that they make considerably less sense than his real analysis books, but sounded quite serious.

"All right, so there's this algorithm for finding digits far out in π. So, what are *you* doing?"

"That's all I'm doing! I'm using the algorithm to find a sequence of digits farther out than anyone has before. I won't know a lot about π, but I'll know something that nobody else does!"

"So, you mean, everyone knows that π starts 3.14 and some crazy people know the next ten thousand digits after that, but you'll know digits one million to one million and sixty!"

"Yeah, that's right, *except* that it only starts 3.14 in base ten, and I'm computing the expansion in hexadecimal!"

"Oh *come* on! Now I know you're really pulling my leg. Why do it in hex? Do you want to read it as an ASCII string and see what it says?"

"No, that's not the only thing hexadecimal numbers are good for. There are some things that make binary and hexadecimal numbers just fundamentally more useful. Like this algorithm; it only works in base two and base sixteen—at least as far as anyone knows!"

Somehow, by its sheer unbelievability, this final explanation convinced Allen that it was really true. Using an algorithm from this unreadable paper, his officemate was about to figure out some previously unobserved digits in the expansion of the ratio of a circle's circumference to its radius. That was pretty cool even if he was forced to do it in base sixteen instead of our usual base ten. Putting down his pencil and closing the notebook in which he was writing his answers to the exam, Allen stood up and walked over to Conor's desk and said "You've got me; that is pretty cool. So, can I watch?"

"Well, cool though it may be, it is not very fast. It's going to take a few days. It'll run in the background and when it's done the digits it finds will be output into a file on my desktop. But, I promise that if you survive your test, you'll be the first one to see what I get."

Conor came into the office and dropped his leather bomber jacket on the beat-up old sofa in front of the blackboard. If all had gone according to plan, the file with *his* digits of π should be waiting for him on his computer. However, while he was rushing over to his workstation, he noticed that Allen was apparently taking a nap on the sofa, with the bomber jacket now covering his feet. He picked up the jacket again and, grabbing onto the rubber toe of Allen's basketball sneakers, began shaking his feet.

"Hey Allen, wake up! Allen. It's ready—today's the day."

"What?" Allen said sleepily. It wasn't clear if he had really heard Conor or if he was talking in his sleep.

"Today's the day, Allen. My computations should be finished. Don't you want to see?"

Allen rolled over onto his side and appeared to stay asleep, but he demonstrated that he understood what was being said to him by his lucid response. "See *what* Conor? It's just going to be a bunch of numbers. If I told you a random string of numbers you wouldn't know the difference."

"Random?!" Conor laughed, throwing his jacket over Allen's head. "But these aren't random. These are *really* digits of pi!" He sat down at the computer and, by shaking the mouse back and forth a few times, got rid of the annoying but colorful screensaver that had filled the screen and replaced it with an image of his desktop. There, under the icon for the office printer, was a new icon representing the output of his computations. He hesitated, afraid to open the file and see nothing but the error message "Error: Division by Zero." So, he was very relieved to open the file and see nothing but a list of hexadecimal numbers:

4C6F6B206174206D652C20436F6E6F

722C206C6F6F6B206174206D652E0A

He had to admit, there was something somewhat anti-climactic about actually seeing the numbers. As Allen had said, although he *knew* they were digits of π, they might as well have come from a random number generator. He stared at the string of base 16 numbers for several minutes, hoping to find something interesting to say about it, but achieved nothing. Then he remembered that each pair of digits in this string corresponds to an English character or punctuation mark according to the ASCII code. Using the string as an input for the program *xunhex* would immediately convert it to its ASCII equivalent so he typed the appropriate commands, but he could not believe his eyes.

The seemingly random string of numbers, when viewed as characters, was the sentence

> *"Look at me, Conor, look at me."*

As he read it, he had the odd sensation that someone was standing behind him, and he turned to see that in fact nobody was there. Still, having read this plea, apparently from the number π to him personally, he was uneasy. His eye began to twitch involuntarily and he felt the urge to run away, which is exactly what he did.

He stood up and walked out of the office door and began pacing in the hall. This may sound strange, but you must understand that as a child, he had always had fantasies that he was really the most important person in the universe, that somehow everything and everyone in the world had been put here for his sake. Of course, as he grew up he recognized these ideas for the self-centered delusions that they were and buried them deep in his subconscious. But now, with this new evidence in hand, they were pushing their way up to his frontal lobe saying "See, I told you so! You *are* the center of the universe."

From the other end of the hallway came a rhetorical question. "Hey, Conor. What's up?" It was David Kooperschmidt, a fifth year grad student who—according to a popular rumor—had written a PhD thesis so amazing that it had already been accepted for publication in *Annals*.

"Dave! Can I talk to you for a bit?" Conor called to his classmate who had just come out of the men's room. "I've got to ask you something."

Dave was used to people asking him questions. He had a well deserved reputation for being a really smart guy, and moreover, since both of his parents were professors at the Courant Institute, he was a polymath with an amazing knowledge of almost every mathematical subject. "Well, that depends," he said smiling. "Do you mean Shannon's information bit or Tukey's ..." Dave cocked his head to the side and stared at Conor's face. "Are you sick or something? You look, like, kind of weird." The question, though not rhetorical this time, went unanswered. "Come on, Conor," Dave continued. "Let's go to my office. I'm supposed to be holding office hours now for the students in my calc section."

Fortunately for Conor, there were no students waiting for Dave at his office. Dave sat down on the sofa that looked just like the one Allen was now sleeping on in the other room, and waited patiently for Conor's question.

"Suppose ..." Conor began, trying to make sense of his own confused thoughts, " ... suppose I told you that there was a message hidden for me in the base 16 expansion of the number π."

"Okay, I'm supposing you told me that."

"Well?"

"Well *what*, Conor? You haven't asked me anything yet."

"Well, what would you think if I told you that."

"I'd probably think you were a delusionary paranoid schizophrenic."

That answer was not really what Conor had been hoping for, so he decided to try again, being more specific.

"Suppose I told you that the string 'Look at me, Conor, look at me.' appears exactly like that—including punctuation and correct capitalization—as a string in the base 16 expansion of π?"

"Now I'd really think you are crazy, Conor. How do *commas* and the letter 'R' appear in the base 16 expanion of π?"

Conor shook his head in frustration and almost shouted "I mean first convert it to character strings using the ASCII code!"

"Oh," Dave thought for a very short moment, shrugged his shoulders and said "Then I'd say that Hardy already knew that a long time ago. At least, he knew it was highly probable!"

"Hardy, the British mathematician? What do you mean? The string said 'Conor' in it ... that's my name!"

"I know your name, dufus. It doesn't matter. Look." Dave walked over to the book case and ran his finger along the spines of the books until he came to the one he was looking for. "Look here. Hardy showed that it is very rare to find a number that does not contain *every* possible finite string of digits. I mean, if you pick a number at random, then with almost 100% certainty, it is an irrational number and also with almost 100% certainty its decimal expansion contains my phone number, your social security number, the entire Gettysburg address in ASCII, and the statement 'Look at me, Conor, look at me.' so I think that it probably is true that the number π contains that expression—and so my diagnosis changes. You aren't schizophrenic, just paranoid."

"No, Dave, you don't understand ...."

His desperate explanation was interrupted by a knock on Dave's open office door. In the doorway stood a student with a skateboard in one hand and a calculus textbook in the other. "Dave," he said "it's your office hours right? Can you explain what the professor said today ... I was completely blown away."

"Yeah, come on in. Take a seat where you can see the board and I'll explain it in really simple terms." Turning to his classmate and holding out the book, he said "Sorry, Conor. We'll have to continue this later. Why don't you take the book so you can see what I mean."

Conor took the book and stormed away furiously. Forgetting about the differential topology class he would miss if he didn't stay at school, he left the building and started heading home. On the way, he stopped to play a video game, just to help him relax and think clearly. Conor's need for an "occasional" video game is very much like an alcoholic's need for an "occasional drink." It was three hours later when he left the video arcade, no more relaxed nor less confused than he was when he had come in. By the time he got home, it was so dark that he had trouble getting his key into the lock at the front door to his apartment building.

He had barely slept at all the night before. Instead, through psychosomatic bouts of vertigo, he pondered the deeper meaning of the events of the day. Does the decimal expansion of π *really* contain every finite sequence? If so, does that make his remarkable find any less interesting? After all, even if every string is *somewhere* in the infinite expansion of π, the amazing thing is that he just happened to stumble upon it.

Then, at about four o'clock in the morning, he had a realization that made everything completely clear. During the following three hours he slept like a baby. Now, he was sitting in his office, explaining the details of his remarkable discovery to his officemate.

"Allen, I've stumbled onto something really important here. As soon as I publish this result, the job offers will just start pouring in!"

"What, because your name appears in an important transcendental constant? I think that ..."

"No, not that! Dave convinced me that nobody would find that interesting at all because, at least as far as anyone knows, *every* finite sequence appears in π's expansion, but I figured out how to show that isn't true. Here, look at this."

Allen looked briefly at several pages of computations in Conor's sloppy handwriting. There were lots of infinite sums, Greek letters and inequalities, but nothing that gave him any idea of what he was looking at.

"Okay," he said returning the notes to Conor, "I give up ... what does it say?"

"It says that I'm completely brilliant! You see, everyone seems to think that there is nothing you can say about the next digit of π from knowing any finite number of preceding digits, but I found a counterexample as a consequence of one of Borwein's formulas. This representation of π ..." he pointed to an equation in his notes expressing π as an infinite series "... seems at first to agree with that idea. However, if you make exactly the right assumption about the first sixteen digits in a substring then you can get a recursive formula telling you the next sixteen!"

Wide eyed, Allen took the scrawled pages back from Conor and said "That would be amazing, Conor. Did you *really* prove that? I mean, not that I don't think you are smart, but lots of really smart people before you must have looked at this."

"I checked it, and I'm really sure. I probably wouldn't have found it if not for that 'message' telling me to look ... but after that how could I help but think seriously about this question."

"Conor, my buddy, you're going to be famous! You're going to win the frickin' Fields medal!"

"I'm glad you like it. Let me show you the specifics. The magic coincidence only happens if you consider a certain string that I call 'magic string α.' If that string appears *anywhere* in the expansion of π, then I can tell you what the next sixteen digits *have* to be."

"No way!" Allen muttered quietly, in honest appreciation of the beauty of this new result.

"Yes," Conor beamed, "and the *really* amazing part is what this string of digits, α, looks like when you write it as an ASCII string."

At this, his officemate's expression changed entirely. His facial muscles tightened as he stared seriously into Conor's eyes and asked "What *does* it say?"

"What I've proved here is that if the expression 'Allen' appears anywhere in the decimal expansion of π then the next characters after it must spell 'is a dork!'." Despite their best attempts to look serious, they both were now having trouble preventing themselves from laughing. "So, 'Allen is great!' never appears in the expansion of π, but I have shown that 'Allen is a dork!' appears infinitely many times ... as it *should*!"

When he finished laughing, Allen asked "So, when did you figure it out?"

"At two in the morning, when the gullible part of my brain finally fell asleep, the smart part stayed awake and said 'Do you *really* believe that those digits of π you happened to look at spell your name?' So, um, did the program even compute *anything*?"

"Oh yeah," Allen reassured his officemate while taking a seat in front of the computer. "It had finished doing the computation before I even got the idea for this scam. So, I saved the actual data in an invisible file. Here it is: `.pivalues`. Wanna see it?"

Displayed on the screen, the actual values of π that Conor's computations have discovered looked no more or less real than the fake ones that Allen had replaced them with. The contents of the file were simply:

4920646F6E277420636F6E7461696E2

0616C6C2073657175656E6365732E0A

The two young mathematicians stared at the numbers for a moment, each hoping to find something interesting to say about them. They turned to look at each other, shrugging in unison, when they both recognized what they had to do.

Allen passed the keyboard to Conor who used it to open an *xun-hex* window and loaded `.pivalues` as an input file. Neither of them could believe their eyes when they saw the perfect English sentence that the number π had hidden for them to find. It said:

"I don't contain all sequences."

Conor's jaw dropped. Allen's eyebrows knotted and he sat still and silent with his lips puckered as if he was about to whistle. Then, together, they called out "Dave!!!!"

14

the object

It had been somewhat sunny when Alice boarded the train in Allston, but before she even got up to street level at Kendall Square she could tell it was raining heavily. The wind blew a fine mist of raindrops down on her as she reached the top of the escalator and it poured down on her as she ran from the shelter of the station to her office building across the street.

One of the things she had imagined when she quit school to start her own company was that she would never have to work on days like this again. Since she was her own boss, she *could* stay home. But, today was special. After months of development and weeks of debugging, their program was ready for its first full trial run. Using Alice's algorithm to model atomic dynamics, the program should be able to predict complicated molecular structures. It should be able to determine what a molecule of some substance *looks* like, the arrangement of its constituent atoms, even if no chemist or physicist has previously worked it out. In fact, it should be able to do it even in situations where she knew the chemists and physicists were completely stumped.

It had been very frustrating for her during these past few weeks. After all, she was quite certain that the algorithm would work. However, since she was not able to program the parallel processing computer that was necessary for the computations, she had to recruit help. Her former classmate, Sophia, was doing the computer science side of the project and had been working on the program for nearly a month now, much longer than she had originally predicted. Every time Sophia thought that the program was ready, it either gave some obviously wrong results or failed to run altogether.

But, she thought that today would be different. After all, the program had worked correctly on a few simple examples they had given it the day before. And, they had given it a more challenging problem—to find a new stable crystalline configuration for a set of atoms with relatively complicated valences—to work on over night and it was still running when she had last checked.

With her bangs plastered to her forehead, her vision blurred by the droplets falling across her glasses, and her wet shoes squeaking across the tile floor, Alice made her way to the elevator that would take her up to the office on the fourth floor. The old elevator shook as it began its ascent. By the time she had reached her destination, a large puddle had formed around her on the floor of the elevator.

Alice knew by the umbrella beside the office door that Sophia had already arrived and that she would look disapprovingly upon Alice's state. Even Sophia's famous condescending glance, that the sopping wet t-shirt clinging to her back and shoulders was sure to generate, could not dissuade Alice from rushing back to the main computer console to see if the final results had been reported.

Sophia looked quickly over her shoulder to see that it was only Alice who had arrived, looked back at the computer screen, and then slowly turned to give Alice a long stare accompanied by a knowing frown. "How can you be such a genius when it comes to math and physics and stuff and such a bozo at everything else?" she asked. "It's called an umbrella ... ever consider getting one?"

Alice began to explain that it had not looked like rain when she left her apartment earlier in the morning, but stopped when she realized that she may very well have ignored obvious rain clouds. And, besides which, this was not what she was interested in talking about. "Did the program stop?" she asked her.

Sophia's intentionally vague answer was a simple "Yes."

"Okay," Alice continued, "did it crash or something?"

"Nope, no, uh uh. No crash or anything."

"So, did it find the minimizing configuration?!?"

"It sure thinks it did." Sophia moved aside so that Alice could see the computer screen. "According to the log file, this configuration definitely has less energy than just about anything else. Not to mention the fact that it looks kind of cool. Sort of like one of those 'bucky-ball' things. Just a simple arrangement spread out into a clean shape ... and the computer already checked that nobody else seems to have identified this configuration before."

"But," Alice said, a bit worried, "it looks like a regular polyhedron."

"Come again?"

"Like, you know what a polygon is?" Alice never started sentences with the word "like" unless she was speaking to Sophia. "A square, a pentagon, a triangle ... like that?"

"Yeah, and a *regular* polygon is one in which all of the sides are the same lengths and all of the angles are the same size. So, all triangles are polygons, but only the equilateral ones are regular. Well, a regular polyhedron is like the same thing except it has *faces* each of which is a copy of the same regular polygon and again all of the angles are the same. You know, like a cube has six faces, each a square, and all the angles are the same?"

"All right, that makes sense to me. So, why do you look like that's a bad thing?"

"Well, there can't be a regular polyhedron with 37 vertices. How many faces does this thing have?"

Sophia pulled the keyboard over to her side of the desk and made a few inquiries. "Well, it's got 32 faces and they all look exactly like this triangle. So, that makes it a regular polyhedron according to what you told me."

Alice threw Sophia a look of disapproval similar to the one that she had received earlier. "But there *is* no regular polyhedron like that!"

"What, you've got them all memorized?"

"Damn right, I do," Alice said, spraying water around the computer room as she gesticulated. "There are only five of them. Shit ... the program's still buggy."

"How many polygons are there?" asked Sophia looking puzzled.

"You mean *regular* polygons? There is a regular polygon with n sides for *any* positive integer n."

"Yeah, that's what I thought. So, am I dumb or something? I don't see why there are only five regular polyhedrons."

"Well, it's 'polyhedra' not 'polyhedrons,' and no, you're not dumb. I guess lots of smart people didn't know it, back in the golden days of Greece. But, it's not so hard to prove. I think Euler did it. He just noticed that the number of edges, vertices and faces always satisfies a certain equation. Then you can check, given a few other simple formulas, that there are only five integer solutions ... and this ain't one."

"So," Sophia tapped the screen with her fingers, "*this* doesn't exist?"

"That's right. Obviously your program thinks it does, but it's apparently wrong. You debug it ... I'm going home to clean up."

"Wait a minute," said Sophia rolling her chair back to block the doorway. "I'm pretty sure my program is working perfectly. Maybe it's your algorithm that's not working ... or maybe you're just wrong about this. Except for your complaint that it doesn't exist, everything about this configuration seems great."

Alice shut her eyes and squeezed the bridge of her nose between her right thumb and index finger. Somehow this helped her to concentrate on the problem at hand, to ignore her fears that this business would be a failure and her overwhelming desire to prove to her friends and parents that she had made a good decision.

"Okay," she said, "so why don't you just 'print out' a model of it. Then we can see if this shape really exists."

"Yeah," Sophia said cheerfully. "I never turn down an opportunity to play with that toy." And she rolled herself back to the computer keyboard.

As she finished entering the instructions, the large machine on the table beside the desk came to life, whirring and flashing. The company's most decadent purchase, but one that they had been assured would be necessary to attract business from the hundreds of biotech firms in the area, this was a device that was able to produce a 3-dimensional plastic model of any object in the computer's memory. Essentially, the object was fused together by lasers firing into a small vat of liquid plastic. Though they had not yet had any opportunity to use it professionally, Alice and Sophia had been using it for everything from jewelry to practical jokes and had begun thinking of it as just another 'laser printer' on the computer.

The instructions had clearly stated that the lid on the machine should not be opened until the green 'ready' LED was lit on the front panel. However, they had learned from experience that they could actually open it as soon as the whirring stopped without causing any problems. In fact, although it was still a bit too hot to touch, the model was essentially already done at this point.

Looking at it as closely as she could from every available angle, Alice had to admit that it certainly looked as if it was a regular polyhedron built out of triangles.

"What could be wrong?" she thought to herself. "The bottom might not look the same as the top. I haven't seen the bottom yet. Or perhaps the lengths or angles are just slightly different from each other, a difference so small that we can't see it and the computer can't differentiate it? Or, I guess maybe I'm just wrong about Euler's formula?"

Alice turned to the cabinet to try to find a piece of paper and pencil on which to recheck her mental calculations and was startled by a horrifying noise, somewhere between a human scream and a screech of tires. Turning back she caught a quick glimpse of Sophia lifting the small model, apparently to look at the bottom, before she apparently vanished into thin air. The screeching sound disappeared as well, leaving only the sound of the model bouncing once and then rolling to a stop in the metal tray in which it was created.

Her first reaction was to grab the model, to try to save her partner. And so her hand shot out towards it, only diverting at the last moment to close the lid on it instead.

"What the hell was that?" she said out loud. There was no doubt in her mind that she had seen Sophia holding *the object* just before disappearing, but just to make absolutely sure she looked around the room, behind cabinets, under tables. Although she realized that something scary was going on, instead of being afraid she felt a pleasant bewilderment similar to the way she used to feel while reading Lewis Carroll. Was it becoming real now? *Alice* in wonderland?

On the large LCD screen, the computer image of the object still rotated in virtual space. Sitting at the console, Alice ran several diagnostics on the mathematical subroutines to check for any possible mistakes and also to verify that the object really was as it seemed. Everything checked out. Even though she was certain it was not possible, the computer certainly seemed to "believe" that this was a new regular polyhedron.

Professor Morris began to open the envelope and then froze in fear. This could really make the difference, he realized, between tenure here at MIT or another job search. Several months ago he had submitted a paper to the most prestigious international journal in differential geometry, and in his hand he held the reply from the editor. The best that he could hope for was an unconditional acceptance for the paper. Most likely, there was going to be a report from a referee suggesting some changes after which the article would be accepted. But, there was the chance, however slight, that the referee report was negative—pointing out serious errors or previous results that this merely duplicated—and that would certainly destroy any chances he had of staying in Cambridge.

Before he could open it, he was interrupted by a desperate knock on his office door. Usually such knocks were only heard right before an exam in a lower level class, when a student who had not been working all along realizes at the last minute that they are almost certain to fail. But he was teaching two graduate classes this semester and neither had a test in the near future. Dropping the still sealed envelope on a pile of papers at one corner of his desk, Professor Morris walked over to the door and slowly opened it. There, completely drenched, stood Alice Wu.

"Dr. Morris," she said "I've really got to talk to you."

"Alice? Hi! I didn't know you were still on campus. Pat told me that you had dropped out! Well, come in, come in. Have a seat." He picked a pile of preprints up off of a chair and put them on the floor next to it. Then, thinking the better of it and worrying about them getting wet, he moved the pile to the top of the filing cabinet and took his seat behind the desk. "So, what can I do for you?"

Alice had always been one of his favorite students. Not only was she clearly brilliant and creative, but she had such an infectious interest in math that made even her classmates more attentive and effective. He began to smile, since he was just happy to see her again, but stopped when he saw her gravely serious expression. She sat still and silent.

"Alice," he asked sincerely, "is something wrong?"

"Yeah, I think so. Look, you remember Euler's formula?"

Now he couldn't help but smile—nothing very serious could be wrong if it had to do with Euler's formula. "Of course I know Euler's formula." He picked up the unopened envelope and waved it, though she could have no idea what it contained. "The Euler characteristic is an essential part of my research."

"Well, wouldn't it say that you can't have a regular polyhedron with 37 vertices and triangular faces?"

"Alice, you *know* that's true. That was one of the first things we did in our Methods of Proof class."

"But," she said cautiously, "what would it mean if I *did* find one? Is math wrong?"

"I'm not sure what you mean, but I can promise you that math isn't wrong. You just have to figure out which axioms to apply. I mean, what we proved is that in flat $\mathbb{R}^3$ there are no regular polyhedra like that. But, that's not the only space there is in the mathematical universe. I guess if you want to find an object like that, you can probably find it in some alternative geometry. In fact, that's the sort of thing that I study: low-dimensional spaces with exotic metrics. You know? No? I mean, like in the theory of relativity where space *itself* is curved."

Suddenly she sat up straight and whispered "Relativity?" The most novel feature of her new algorithms was their incorporation of appropriate parts of relativity and quantum physics. She did not really understand either of these theories—did anyone?—but she knew that they were both very close approximations to reality in their regimes and so she took some formulas from each.

"What would it mean", she asked, "if there was one of those objects here in this room, in your *hand*?"

"What, you mean one of those 37 vertex regular polyhedra embedded in a 3-dimensional manifold? I can't say off of the top of my head, but if you give me time I'll try to think about it!"

She jumped up and shook his hand, dripping rainwater onto his desk. "Thanks, thank you," she said, and then she left.

Professor Morris closed the door behind her and picked up the envelope. Opening it carefully and unfolding the contents he read *"Your submission, 'Equivariant Chern Bundles on Spin-3 Manifolds' was positively reviewed and will be acceptable for publication in our journal if an attempt is made to comply to the suggestions of the second reviewer (see attached) ..."*

Alice could not figure out how to feel or what to do. What *was* the sensible thing to do when things don't make sense? She felt panic encroaching whenever she tried to imagine what had happened to Sophia and whether she would ever see her again. At the same time, she felt positive about her interaction with Professor Morris, not only because she missed school but also because she felt that it helped her to understand what had happened. She certainly did not *really* understand what happened, but the existence of the object seemed to imply an alternative geometry. Could this be the consequence of the part of her algorithm that considered slight variations in the space-time metric?

Rather than taking the train home, she decided to walk. The rain, that was beginning to subside anyway, certainly could not get her any more wet than she already was. In Central Square she grabbed a latte and a croissant. In Harvard Square she browsed the periodicals at the news stand. On the bridge she stopped to look at the river slowly flowing beneath her. She had rather successfully distracted herself from her problems, which she knew was precisely what she had to do to make any progress on them. Somehow, it was when her conscious mind was occupied with mundane minutiae that the mathematician in her subconscious was able to make the most spectacular discoveries.

And so it was that as she walked up to the front door of her apartment building, trying to find her key in the dim light of the sunset, she suddenly realized how she could investigate the question by hand despite the complexity of the problem. If she did not know anything about the final form of the object, she would never be able to figure it out without the help of the computer, but knowing the symmetry structure of the object that the algorithm found in the end would allow her to investigate it herself.

She ran up the beat up marble stairs, the sound of her steps echoing through the building. Turning on the ceiling lamp inside her small apartment revealed the large room that served as her bedroom, living room, office and dining room. Grabbing some scrap paper from the filing cabinet and a pencil from her desk, Alice sat on the futon and proceeded to try to reproduce the object from first principles.

The next morning was sunny and bright. Approaching the office building she found it difficult to believe that anything unusual had happened at all. She expected to see Sophia in the office as usual, insisting that Alice had imagined it all in a dementia brought on by excessive wetness. This pleasant fantasy disappeared when she saw the two police officers waiting outside the office door.

"Good morning, officers," she said, not yet certain how much she was willing to tell them about what really happened, "what can I do for you?"

"Would you be Alice Wu, then?" asked the taller officer.

"Yes, that's right."

"And your business partner is Sophia Yakimov?"

"Yeah, that's right." She fumbled nervously for her keys and motioned to indicate that she wanted to open the door. She was thinking to herself *'don't lie, whatever you do, don't lie ... you'll only be*

making it worse' but she turned to the other officer and said as inno-
cently as she could "Why, has she done something wrong?"

"We've got no reason to believe so. No, we're investigating this
presently as a missing *person case* only, young lady."

"*Missing?*" she said in mock surprise, "I saw her two days ago and
she looked fine then." '*That's true,*' she thought '*I did see her two days
ago.*'

"And that was the last time you saw her?" The other officer
stepped aside so that she could unlock the office door. She undid
two deadbolt locks and swung the door open, holding her breath
in fear that she would see something unusual or scary behind it,
but the office looked the same as ever.

"Yes," she said, while thinking 'NO! DON'T, DON'T LIE!'
"That was the last time I saw her. Come on in. Make yourselves
comfortable."

"So, didn't you find that somewhat ... unusual? That you didn't
see her yesterday I mean."

"Not really. We've sort of got an understanding that we can
show up at work whenever and miss work whenever as long as we're
doing a good job and making progress. For instance, I was out one
day last week when a friend of mine asked me to join him on a hike
up Mount Monadnock. I told her about it when I came into work
the day after. No big deal."

"So, were you here yesterday all day?"

"No," Alice answered, happy for a question she could answer
honestly.

"I was here early, like 8:20 AM or so, but then I was out from ten
on."

"Where did you go?"

"Oh, I met with a math professor at MIT to ask some questions
related to my work here and then I walked home up Mass Ave and
over to Allston around Harvard."

"You *walked* home yesterday? Wasn't it raining rather hard yesterday?!?"

"Yeah, it was and I got soaked. I bet if you ask people in Central
and Harvard if they saw a soaking wet Asian girl yesterday they'll

remember me. I must have been a sight. Um, what makes you think Sophia's missing?"

"Her boyfriend contacted us when she didn't come home from work yesterday. If, as you say, she didn't come in to work yesterday then she's been missing for more than 24 hours now."

"What?! Sophia has a *boyfriend*?!"

"Are you two close friends?"

"Me and Sophia? Nah, just partners. I mean I like her enough, but we don't really talk about much other than our work."

"So, do you have anything else to tell us that could help us with our investigation?" The officer pulled out his business card, but noticed that Alice kept glancing behind them to the metal cover that concealed the object. He tapped a corner of his card on the cover of the chamber, producing a dull *plonk* sound. "What's under here?"

"That's a machine we use in our work. It can make a plastic model out of a computer representation of one."

"Could you open it up for us, please?"

The officers stepped aside and Alice slowly lifted the heavy metal cover. Peering inside when it was only partly open she saw that the object was still there, right where it had landed when Sophia dropped it.

When the cover was almost completely opened, she saw a small shadow move out of the corner of her eye. The shadow appeared to come from the object, fall out of the chamber, down the side of the cabinet and land silently on the floor. Instinctively, she jumped back and inhaled sharply.

Both officers looked alternatively at the chamber, empty apart from the object, and Alice who was clearly frightened. "What is it? Is something missing?"

"No, I thought I saw something fall out of there is all."

"I didn't see anything. Did you see anything fall out of there, Webster?"

"No," the officer stared at her with visible signs of suspicion, "nothing fell out of there."

The two officers left after giving her their business cards. She could hear them mumbling in the hall when the door closed behind them, certain that they were talking about her. She was a terrible liar.

"Professor Morris, got a moment?"

"Alice! I was hoping you'd come by again. I found something sort of interesting." When he walked over to the chalkboard and started erasing formulas to make room for whatever he wanted to show her, she realized that it must be something big. "Now, this may not be at all what you were looking for. I'm not sure *what* you were thinking when you asked about it, but that object you were talking about turns out to be really interesting. Have you ever heard of *instantons*?"

She shook her head slowly from side to side, but said nothing.

"So, what I found is this: that object you found is interesting because I can put it in a space that looks just like flat Euclidean space everywhere except right near the object. The distortion is *localized*."

She nodded, thinking that this made sense and that she might have said the same thing about the problems caused by the object since everything *away* from the object seemed to be completely normal. "Okay, that makes sense to me. So what's an *instanton*?"

"Oh, that's a very famous example of this sort of thing, though not exactly the same. In that case the localization is in time as well as in space, so it is interpreted as an instantaneous effect. It doesn't change the mathematics at all, just a way to think about it. The one I found from your suggestion is more of a topological soliton."

He remembered the smirk she gave to professors when they said things that she did not understand, as if to show that she was disappointed in them for not being able to say things clearly enough that she could follow.

"I guess the way you think about that is that it's a sort of a bridge between two different vacua. Like two different universes that connect at just one point?"

"Two different universes?!?" She sat up straight, eyes wide open.

"Now, don't get carried away with that. Like I said, it's just a mathematical formalism. It doesn't really mean anything. Like this:

the vacuum down here just goes along straight," he illustrates with a horizontal line on the chalk board, "and then in a localized disturbance we call a topological soliton it switches almost instantly to this other vacuum." The line he is drawing bends sharply in the middle going up to another height before continuing horizontally again, apparently forever.

Alice stood up, as if she was going to leave without saying a word, but then turned back and pulled a piece of paper from her back pocket. "Is this the geometry you're talking about?"

Professor Morris took the paper and glanced at it quickly. He was shocked at what he saw because he thought that he was the only person in the world who knew the formulas that were written there.

"Where did you get this? Did you figure this out *yourself*, Alice?"

"Yes, it comes out of my molecular structure algorithm in the right circumstances. The math for finding it was actually pretty simple once I figured out what I was looking for."

"Look," she said absently, "I've got to go. Let me know if you find anything else, okay?" Before he could say anything, she was gone.

The next morning, tired from a sleepless night of computation and worry, Alice came up from the T stop and headed across the street to her office building. She was disturbed to see the two police officers from the previous day standing next to an ambulance by the entrance. Cautiously, she walked up to them, squeezing her arms as if cold, in fear of what she might find out.

"Miss Wu," said the tall one.

"What happened, officer? What did you find?"

"You've got quite a flea problem in your office, ma'am."

"Webster! What are you talking about?"

"They stopped just about the time you arrived, Bill, but when I was in there at first looking at the body I was getting bitten pretty bad, and it still hurts."

"A *body*? There was a body in my office? Was it ... " Alice couldn't finish the sentence. She knew whose body it had to have been. The two policemen looked at each other, not certain whether to discuss this with her as a concerned acquaintance of the victim or as a murder suspect.

They nodded, having reached a silent agreement.

"A cleaning woman found what we presume is Sophia Yakimov's body. She was badly mutilated, beyond recognition. The body was found in your office. Can I ask who, besides yourself and Miss Yakimov, had a key to the office?"

"Her *body*?" Alice repeated. "Where was the body? Was it there the whole time?"

After a quick glance at each other, the smaller officer said quietly "As far as we can tell, it was not there very long when it was found this morning since the blood was still wet and had barely soaked into the carpet." The thought of her partner's bloody mutilated body on the floor of the office was too much for Alice. She grabbed her stomach and fell to the ground while the world seemed to spin rapidly around her.

After quite a bit more questioning, the police finally left Alice sitting alone on a stone bench in front of the office building. They even told her "not to leave town," just like in the movies. Alice knew that she would not be able to open the office door, but she was very, very afraid. She had to do something about this right away and so she headed straight for Professor Morris' office at MIT.

Running through the so-called 'infinite corridor,' she almost knocked down a student walking on crutches. She could not get enough breath while running up the stairs to the math offices, but kept running despite the pain. When she got to his office, Alice sat on the floor breathing heavily for several minutes. People walking past tossed her strange looks and made an effort to stay as far away from her as possible so she realized that she must look crazy. And she *was* crazed; she pounded on the professor's door repeatedly, even though she knew when he did not answer right away that he must not be there.

As soon as she was able, she ran back to the stairwell and headed up towards the math office, where she hoped she could find

Professor Morris. She literally ran into him between floors. Unable to speak for lack of breath, she just put her arms around him and hugged him.

I don't know if this makes sense to ask you," she said once they got back to his office, "but is there some way to destroy the object? I mean, can we mathematically eliminate it or something?"

"Alice, it's amazing how you are always one step ahead of me on this. You really ought to consider coming back to school. Your talents are really ... "

"Professor, please!"

"Right, sorry. I see that this is urgent ... somehow. Anyway, that is exactly the question I was working on last night, though I wouldn't have put it that way.

"You remember," he continued, pointing at the illustrations that were still on the chalkboard from her previous visit, "how I viewed the geometry surrounding your object as a bridge between this vacuum down here and this other one up here? It has an *orientation* because you can see that it goes from down-left to up-right. So, a natural question is 'What about a bridge with the other orientation?' I looked into it and it turns out to be some really pretty math."

"Pretty, in what sense?"

"Well, you remember your object."

"Yes, I can't get myself to forget it."

"It has a natural dual ... a similar object made by exchanging vertices for edges and vice versa. Interestingly, the dual object can only exist in the geometry corresponding to this reverse oriented bridge! Of course, you've got to understand that these are just one-dimensional diagrams that I'm using to describe the ... "

"Professor! What does this have to do with my question?"

"OH! Don't you see? The bridge one way is a soliton and the bridge in the other direction is what we would call an anti-soliton. When they collide, just like a particle and its anti-particle, they just cancel each other out."

She nodded and smiled. It made sense to her, and it also seemed as if it might work. Before he could say "Alice, can you tell me what is going on here?" she was gone.

Under her arm she held several pages of computations that she had done the night before. Although she didn't know anything about solitons and anti-solitons, she was pretty certain that she had found the appropriate geometry. It seemed to be dual to the one she had found earlier in the sense that it was almost the same, but somehow inverted, and it supported the dual object with edges and vertices interchanged. Now if she could just get to the computer and get it to 'print one out' ....

The image that greeted her was worse than she had imagined. Through the yellow "Crime Scene" tape, she could see the splattered blood on the walls and the huge stain on carpet near the door where the body must have been found. In the corners of the room most distant from her, she could see the two things that had brought her here. The computer, running a screensaver generating fractal images, ran in one corner of the room and the 'printer' with its chamber presumably still containing *the object* sat silently in the other corner. After checking to make certain that nobody in the hall was watching, she pushed aside the tape and entered her office.

It was not until after she sat down at the computer that she remembered the shadow she had seen 'falling' from the object. She looked carefully around the room and was terrified to notice not just one but close to ten tiny shadows on the floor, cabinets and tabletops. Their size changed, depending on their location relative to the object. The closer they got to the chamber the larger they became, so that when one was on the cover itself it seemed to be the size of a fist, while closer to her they became the size of a dime or even smaller.

Realizing that there was no time to waste, Alice began typing as quickly as possible, trying to describe the dual object to the computer in precise mathematical terms. She was not certain at all that she would be able to understand Sophia's code, or that she could bypass most of her own algorithm and simply get the computer to recognize the new object she wanted it to make. The task was made even harder when she recognized that two of the shadows, now barely visible since they were the size of pinheads, were 'crawling' on her legs.

The 'bites' of the tiny shadows were extremely painful, and the blood running down her leg seemed to attract the other shadows that began moving towards her. As soon as she had instructed the computer to begin 'printing' a solid model of the new object, she jumped up and ran as far as she could from the object without leaving the room. This strategy seemed to work. The shadows attempting to follow her became far too small to see and the ones on her legs seemed unable to move as quickly or do as much damage when they were so far from the object. Looking down, she noticed she was now almost exactly on top of the large blood stain. Had Sophia come to this same spot in the room with the same idea in mind? If so, it certainly had not worked for her.

The machine stopped making noise, indicating that the new object had been printed, but Alice noticed no change in the situation. The tiny shadows continued to bite her legs and she did not notice any change that would lead her to think that she had succeeded in destroying the original object. Once the machine signaled that the object was completely done, Alice knew that she would have to open the cover on the chamber to see what she had actually accomplished.

In the brief time that it took her to get back across the room, the shadows on her legs grew significantly larger and were able to inflict much more serious damage. The other shadows, still trying to get back from the far side of the room, were also getting larger *and* faster. She lifted the heavy lid to reveal the two objects—the original and the new dual object that she had just created—sitting near each other in the chamber. It was her intention to pick up the new object and somehow force it and the old object together, but she did not have the chance. Before she could do anything, the two objects flew together as if compelled by a magnetic force and—accompanied by a loud popping noise—became a single, ordinary triangular pyramid. The shadows on her legs, and those on the floor that were just about to reach her, disappeared.

Again, she fell to the floor, clutching at the wounds on her legs to stop the bleeding. As she sat there, the computer screen began to display another screensaver—this one alternated floating images of the company logo with pictures of Alice and Sophia goofing off around the office—and Alice began to cry.

15

The Legend of Howard Thrush

Back about half a century ago when I was a youngster like you, things were different. The horseless carriage hadn't yet invaded our country's roads and highways. The Wild West really was wild. And facts were more true than they are today. Sure, you'll hear plenty a tall tale about people from those magical days. About Pecos Bill and Johnny Appleseed, Paul Bunyan and John Henry I can't say, but I'll tell you one thing: those stories about Howard Thrush are for real. I know 'cause I was there, a witness to them all. In fact, Howie and I were kids together, and his greatness was plain as the hypotenuse on a right triangle even then.

I wasn't much of a math whiz myself. When Miss Hagerty the school teacher got around to teaching us arithmetic, we were hearing all about some lucky kids who had ten pennies and how three got taken away. Then she'd go and ask us how many pennies was left. Being dimwitted as I am, I was the first to answer—that ten were left but that someone else had gone and took three of them— and got rewarded with a good hard slap on the forearm from her pointing stick.

Most of the other kids in the class probably knew I was right, but they seen what she did with the stick and knew that she wanted them to say that only seven was left, and that's what they said.

But Howie, he was a regular genius already. He'd seen how angry she was when I said that there was still 10 and he knew then she had something to hide. So, he started in wondering where all those things that got taken away in arithmetic had got to. Not just pennies mind you, though we frontier kids would've been happy enough to get our hands on those, but apples, carrots, roofing nails, and best of all, candy bars that we never even seen in our lives but only read about in stories about these kids who didn't even seem to mind them disappearing all the time.

So, one night, Howie sat up calculating real slow. It was hard for him, because he was usually fast as a yam dropped out of a hot air balloon when it came to doing anything with numbers. No, he wanted to do it real slow like so that he could see what really happened when these things got taken away. Then, like a spider crawling on the wall that you only spot out of the corner of your eye, he noticed them: a secret stash of numbers that Hagerty'd never told us about! Not one and two or even the fractions that we'd started in reading of late, but a different sort of numbers, shy ones that hid behind the "take away" sign so as we wouldn't notice them. But now he'd seen them and they couldn't hide anymore.

Now that we knew where she was storing away all of those things that had been taken away over all of those years of teaching arithmetic to us kids, we had quite a party with the apples and pennies and carrots and roofing nails and especially the candy bars. There couldn't have been much left of her stockpile when we were all done, but the next day at school, out of embarrassment and shame, she didn't even mention it.

From that day on, I knew that Howard Thrush was destined for great things, though I was still gonna be surprised at exactly how great. When I was out of school and working with the blacksmith all day, I'd go talk with Howie about what he'd been doing when he got home from class. His teachers were, as you might expect, always impressed with that young feller. In geometry, he got attention by being able to trisect an angle using a compass and a straightedge, even when the edge weren't quite so straight. His algebra teacher had something to write home about when Howie showed him a new way to extract roots of a polynomial two at a time, a lot faster than the old-fashioned way. And he saved his trigonometry teacher from using up all his chalk by pointing out that all of the functions he liked to talk about could be called by just the first three letters of their names instead of writing them out in full all of the time.

But, of course, it wasn't until '89 that he went and did something that made him the legend you all know of today. It was a bad year for mathematicians, the worst ever according to some. Not the ideal time for Howie to be finishing up his PhD at the less than ripe age of 17, but that he was.

When he first went into grad school, it looked like it was going to be quite the golden age for people like him. Mathematics was, rightfully, becoming more and more important in the world. People were hiring mathematicians to help them with everything: designing flying machines, figuring out travel plans for salesmen who wanted to visit a bunch of towns once each, working out how much to charge per yard for their carpets to maximize their profits and how much paint to buy if they needed to cover the area under the curve $y = 1/x^2$ between $x = 0$ and $x = 1$.

The problem was, with so many mathematicians working on problems day and night, there just weren't enough *variables* to go around. It didn't take long for them to notice that with just a to z and A to Z, there was barely enough to last out the year. Talk of the *Great Variable Shortage* filled the mathematical mess halls and all those great brains set themselves to figuring out what to do. The ones who'd spent too much time in civil work proposed a rationing, and seeing as nobody had a better idea, they started in with that. At night, when few people were working anyway, you could use just about any variable you wanted, but during the day only people working on the most important of problems had access to the choicest of them like x and t.

The few mathematicians with a good liberal training in the classics made an important contribution by suggesting the use of foreign substitutes like θ and $\aleph$, but everyone knew that this was just delaying the inevitable. And nobody liked to think what would happen then.

This is when Howard Thrush made his grand entrance, still wet behind his ears and having him a special iron slide rule that I'd made myself. He was all set to solve this disaster and make a name for himself, though I must humbly admit I had a bit part to play in this legend too. You see, even when he'd been working on his thesis in the days before the GVS, he was thinking how he liked numbers better than variables because he didn't have to come up with names for them. I mean, you don't have to name the number two, it's just *two*, and even good old minus the square root of twenty-nine comes to you prelabelled, no naming necessary. On the other hand, when he had to come up with variables he always had to waste time thinking of what to name them all, and worrying which ones were not used up yet. He figured that if he could eliminate this problem he'd save so much time he could double his productivity, mathematical-researchwise. What could he do though? He didn't think about this again until '89 when it had become all much more serious.

At this time, I was working on a real big project for that crazy sea captain who lives on the long pier by the lake. He wanted me to build him an ironclad boat that could go underwater—a "sub" he called it. Days and weeks I spent slaving at the bellows and anvil making bent rectangles according to his fancy design. They all looked the same to me, and so to tell them apart I started putting numbers on each one. A one on the first, and two on the second. I could see what Howie meant ... it was nice not to have to think of any names. So, I suggested that he do the same. "Why not just use one and two as variables?" I said to him. "Just name 'em in order as they come up."

He said he couldn't do that, and he's probably right though I still don't really understand why. There was a difference between variables and sub parts, he told me. Then, in a flash of genius he saw the solution! He used the idea in his thesis right then, instead of naming each variable individually, calling them all x, but telling them apart with *sub* numbers, like x sub one and x sub fifty.

His math professors were so busy worrying about the GVS and trying to beg steal or scrape enough variables together to get their work done that they didn't even realize Howie'd solved it all. He finally showed them that using sub numbers meant that there were more than enough variables for everybody, infinitely many in fact. When other mathematicians heard the news, they came flooding into our little frontier town of Princeton, NJ and what ensued was one hell of a mathematical smorgasbord that lasted for near ten months!

There were seminars all day and colloquia all night. They was so gosh-darned happy that even their failures got turned into successes. For instance, one night the geometers had more than a bit too much coffee to drink. They were so fidgety that they couldn't even draw a line parallel to a given line, poor folks. At another time, they would probably have decided to just call it quits for the day, but at the smorgasbord these guys were as wild as a nowhere-differentiable-function. Instead, of giving up, they just gave it a name, projective geometry, and went on doing geometry without any parallel lines.

But, by the end of ten months things were getting out of hand, and Thrush he knew it too. The town was so full of mathematicians that for any theorem proved on one side of town you could be sure to find someone on the other proving the opposite. So, Howie got together a team of a few hundred or so mathematicians and support staff and they headed West.

Now, you might think that Howie'd choose just the best mathematicians, right? He certainly could have, but no. He chose who to take with him based on how friendly they were, figuring that personality would be more important than genius ability on a trip like that. And quite a friendly bunch they was too—at the start.

I can't say as I met *everyone*, but the folks I got to know were awful nice and smart. There were the topology twins—Betsy and Barbara—who not only could find their way around any non-orientable surface, but they also were good enough at packing problems to get all of the food for the journey into a single wagon. Cheerful Mike McGillicuddy tended the horses and organized the seminar schedule. We also had quite a quorum of fellers from the Far East with us. Most of these folks were experts in micro-local analysis, but just a few of them were along for no reason other than cooking up a good chop suey every once in a while. And, of course, Howie was kind enough to invite me along even though I couldn't do nothing but smithing.

The trip started out pretty uneventful, sad to say. Still, we had a great time, seeing the parts of this beautiful country and goofing around. One night, for instance, we had a contest of joke telling. But, of course, there was a special mathematical twist to it. Every joke told had to end with the punchline "Oh, I'm sorry ma'am. I thought you said 'cohomology'!" You can imagine how much fun *that* was.

Then, when we were going through the state of Indiana, who did we run into but Colonel Bill Cody and his Wild West Show. Thrush didn't want to go, but the rest of us begged and not only did he wind up taking us, but he winded up being the star of the show!

It was Howard Thrush himself who got called up to challenge Miss Annie Oakley when it came to that part of her sharpshooting act. She showed us her shiny, silver rifle and told Howie that he could choose his own weapon. The crowd just went crazy when he told them he was going to use the natural logarithm function, ln x. It sure seems impressive. I mean, there's a vertical asymptote on the graph, the domain is just chopped off to *half* of the real number line, it's got something to do with the mysterious number e. And, worst of all, it grows so slowly that it looks like it's never going to get anywhere.

But Howie knew what he was doing. All of that fancy stuff just distracted people from noticing one important thing: the function's *range* is unlimited! So, of course, Howie could hit *anything* with it! Annie would hit a few targets, and so would Howie. Annie switched to using a mirror, and Howie pulled out the exponential function. When Annie shot two or three targets at once, Howie started using vector-valued functions. For an hour, the crowd watched the two match each other at every challenge until it was finally called a draw.

We left the show carrying Thrush on our shoulders, feeling great. We didn't know that he'd put himself into a lot of danger.

You see, the next morning he was kidnapped by a bunch of rustlers who saw him at the show and figured he'd be able to bring in a pretty penny in ransom.

Of course, I didn't know that at first. I only knew he was missing from camp when we woke up in the morning. I was real worried, but the others told me not to be. They pointed out that Howie was smart enough to get out of any kind of trouble. I should have known they were right ... and so should those rustlers.

When he got back, he told us what happened. First, those nasty fellers drug him out of his bed, *and* drugged him out of his consciousness too. He woke up with his hands tied behind his back with a good, strong rope, sitting on a chair in the middle of a room with thick, cement walls and no windows. Through the wooden door, he could hear the rustlers talking about what they'll do with the ransom money.

What could he do?

Well, the first thing he did was a proof. He showed that if $f(z) = z + a^2z^2 + a^3z^3 + \cdots$ is univalent in the unit disc then $|a^2| \leq 2$. The point is, as we all know *now*, this is a *sharp* inequality! So, he used it to cut the rope.

Now, the room was practically empty. All that was in it was a big, old desk. And all that was in the desk, in the back corner of one drawer, was an old, broken pencil. But, of course, in the hands of Howard Thrush, a pencil is a powerful tool.

What he did, quietly so as the rustlers wouldn't catch on, is to triangulate the walls, ceiling and floor of the room (standing on the desk when necessary). He just covered that room with a thousand triangles. But, you see, he did it in a clever way so that there were exactly 1502 vertices, 2504 edges and 1000 triangle faces. Then, since 1502 − 2504 + 1000 = −2, this made for two *holes*! One of them turned out to be too small, but the other was just big enough that he could squeeze through it and get back here to safety.

With our leader returned, we started on our way again. But, by this point, everybody was getting on everyone's nerves. The different subdisciplines were at it like the horns of Alexander's sphere. The number theorists couldn't understand why the geometers insisted on working over $\mathbb{R}$ and $\mathbb{C}$ when the smaller fields were "so much more interesting." The geometers hated the topologists for the way they couldn't care less about the metric. The logicians kept annoying everybody by suggesting that there was no rational foundation to what any of the others were doing anyway. And so on.

Just in time, we reached our destination. Right across the bay from the boomtown of San Francisco, Howie decided to set up shop. And that was may be his best decision ever. You see, those poor San Franciscans—out there for so many years without an academic to be found—had so many mathematical questions waiting to be answered, the newly arrived theoreticians barely had time to take a breath between their proofs and computations.

When someone would come in for some math help, McGillicuddy led them through the triage. "A question about the boundaries of a claim? Geometry in the Red Wing. The fractal dimension of the Rockies? The complex dynamicists just down the hall can help you. Think there is an inconsistency in the town's constitution? Wouldn't be a surprise, but the logicians up in the tower can resolve it for you. A ladder whose base is sliding away from a wall at a constant speed? That would be related rates ... just follow the red arrows on the floor. And while you're at it, why not stop in to see the topology twins about that problem with the donuts and the coffee mugs?" And so on.

It seemed to all of us that these golden days—and with gold nuggets as the currency of choice out there, I really do mean golden—would go on forever. The questions sure seemed endless, and the combination of brainpower they had working on them showed no signs of diminishing either. For years we all stayed there, working under Howie's wise leadership on any problems that were brought to us or just caught anyone's fancy. But of course, no good thing can go on forever.

One day, when he got a bit distracted and accidentally broke the law by dividing by zero, Howie got it into his head that it was time for him to retire. So he headed up North, towards Victoria Island, BC, all by his lonesome. The popular version of this yarn, that you all probably heard a hundred times, says that nobody knows for sure if he made it there alive. Don't you believe it! All legends, even the true ones, take some liberties here and there to try and sound real dramatic.

Of course he's there! And retirement for him doesn't mean staying clear of mathematics neither! I heard from him just the other day. It's hard to say for sure from those short, choppy wire messages, but he sounded happy to me. Says that just since last Tuesday, he's proved Fermat's Last Theorem, figured out if P is the same as NP, resolved the Riemann and Continuum Hypotheses and he plans to prove the consistency of all of mathematics by summertime. And, believe you me, that's just for starters. I wouldn't be at all surprised if Howard Thrush didn't reappear and shake-up the mathematical world again sometime soon.

16

Progress

Part 1: Year 1918, Oxford University

James Mathewson, a second year maths graduate student, knocked cautiously on Professor Wilson's office door. He was not certain why a professor of the Department of Archaeology would ask him to a meeting, but he was afraid it might not be a reason he would like.

When he came in, Professor Wilson began looking through the folders on his desk, looking for one that was apparently relevant to the discussion with Mathewson. He found the folder and leaned back in his seat.

"Mr. Mathewson, thank you very much for agreeing to meet with me. I believe you may be able to help me."

As the professor flipped through the papers in one folder with his left hand, his right hand holding the folder was shaking so violently that the papers rattled like distant thunder. Mathewson wondered about the professor's tremors, thinking it unlikely that they were an indication that he was nervous about the present meeting.

"I help you? Why I would be very happy to do that ... but how?"

"There has been a tremendous amount of activity in the study of the ancient Egyptians as a result of our new ability to read their hieroglyphic writings. Before this discovery, we had very little idea of what life was like in Egypt, what the people there knew, how they thought. Do you mind if I smoke? No?" Professor Wilson packed some sweet smelling tobacco into the bowl of his pipe and lit it with a small silver lighter.

"I have read a bit about ancient Egypt, fascinating stuff that. I'm afraid I cannot say that I know of any of your own work, though, Professor Wilson. What is it that you do?"

Though this struck him as somewhat impertinent, Professor Wilson simply answered the question. He decided that this was just an example of the blunt, 'no nonsense' style for which mathematicians are known. "I enjoy looking at the frontiers of knowledge in ancient cultures that we would otherwise consider to be *primitive*. For instance, the Egyptians have built those magnificent pyramids. Surely that must have involved an advanced knowledge of engineering, beyond what we would normally attribute to a pagan society."

Wilson now began looking through the folder in his hand. The confidence with which he spoke and his otherwise calm demeanor convinced Mathewson that the professor was no more afraid of him than a python fears its prey. Consequently, he began to have pity upon the poor old man for the affliction that must trouble him continuously. But, his attention was drawn back to the question at hand by the man's intense stare and deliberate speech.

"In fact," the professor continued, "one might infer from the dimensions of the Great Pyramid of Giza that the Egyptian clerics knew about the mystical mathematical number we call π many thousands of years before Archimedes."

"Aha!" Mathewson said, in outburst so sudden it created a visible disturbance in the clouds of grey smoke over the desk. "Now I see where you are going, professor. You would like me to tell you what sort of mathematics they would have needed to ..."

"No, not quite. I have a much more specific question for you to consider." Wilson handed Mathewson a page of incomprehensible scribbles. "This is the work of an Egyptian mathematician from over 3000 years ago. It is a copy, of course, don't worry about damaging it! I know precisely what this man intends to do, he explains it here at the top where he poses a question, but then I fail to understand how this collection of numbers here in the middle leads him to the answer." Wilson used the rapidly vibrating tip of his pipe to point to the center of the page, where Mathewson now recognized sets of parallel lines grouped into sets of size less than ten as a primitive way to write numbers. "I approached Professor Hilton in your department, asking for his help, and he suggested that one of his young students—you, in fact—might be interested in such a puzzle."

"Well, he was certainly correct, it sounds like a puzzle that I might enjoy, and I suspect that I can help you with it. Please, if you can show me just a bit more of the notation and then let me borrow a copy of this manuscript, I should be able to explain the computation to you tomorrow at this same time."

"That would be wonderful! But I warn you, it may not be as easy as you imagine. I know my arithmetic quite well, but I cannot see at all how these numbers come into the computation of 13 × 7, which is all he is meant to be doing here. Why, if I was to multiply 13 × 7 now, I would be able to do it in just three lines, and I would end up with the same 91 as this man did so long ago. However, here in this column he has written 1, 2, 4 and 8," Mathewson observed that Wilson pointed to a single line, a pair of parallel lines, four lines in a row and finally an array of eight lines in two equal rows concluding that his interpretation of the symbols was correct, "which have absolutely nothing to do with this computation!"

"So, this then is the symbol for ten?" Mathewson pointed to a small horseshoe shaped symbol.

"Yes! Yes it is. I did not know that you knew hieroglyphics already. This will be easier for you than I had imagined."

"You misjudge me! I did not know a single hieroglyph before I came into this room. But, I do know mathematics and that is the same in any culture. So, am I right when I say that this here," he pointed to three spirals directly to the left of a pair of parallel lines, "is three hundred two?"

"I see I have asked the right man! Wonderful, wonderful, wonderful." Wilson handed Mathewson a yellow sheet of legal paper from his folder and then, thinking the better of it, simply gave him the entire folder. "This sheet contains a brief dictionary of the mathematical notations I have been able to discern so far. These numbers, these almost identical 'walking legs' for addition and subtraction and so on. Why don't I leave this with you, as well as the transcription of the entire papyrus, and we can meet back here tomorrow as you suggested."

Mathewson took the folder and, in complete silence, looked at each page individually. The professor sat across from him, also in silence, waiting for the student to agree to the suggestion, or to ask for clarification. When he had finished, Mathewson stood up and left the room without saying a word. Wilson shrugged and returned to the tablet that he hoped would shed some light on the ancients' understanding of the flooding of the Nile.

Part 2: The Following Day

"These Egyptians were much more primitive mathematically than you have lead me to believe, Professor Wilson, pyramid builders or not."

This is not at all what he had hoped to hear from Mathewson, but you would not know this from his reaction. He said simply "And what makes you say that?"

"Do you see what they are doing here?" Mathewson passed the professor a page of transcription from the mathematical papyrus. In fact, this was the very page that the professor had originally tried to understand. However, because of the difficulty of the mathematics, he had given up and attempted to read some more elementary computations.

"This is a very interesting page, indeed," agreed Professor Wilson. "I know what he is doing, he is showing how one can resurvey property lines after the floods of the Nile have receded."

"Oh," the student said, looking at the page again with a renewed understanding. "I did not realize the particular application ... In any case, the geometrical significance was clear. In fact, with nothing more than first year geometry this would have been entirely elementary, but it takes your 'scribe' several pages to work it out. I think they could have done much more with mathematics, and then much more with science and engineering as well, if only their method of *multiplication* was not so cumbersome and inefficient."

"So, now we come to the question I posed to you yesterday, and I suppose from what you have just said that you can answer my question. How is it that the ancient Egyptians multiplied numbers?"

"The short answer to your question is that they did it in an accurate, but completely awkward and primitive fashion. Suppose, as we considered yesterday, that our ancient accountant wished to multiply 13 by 7. He would start by writing out a list of the powers of two in a column." He demonstrated the notation for the professor, using Arabic numerals, since they were still easier for him than were the Egyptian hieroglyphs:

$$1$$
$$2$$
$$4$$
$$8$$

"They stop the list before they get to a power of two larger than the first number—just as I've stopped here before getting to 16 which is larger than 13."

"Already I agree with you that their methods were bizarre, since none of these numbers you have written down would even fleetingly cross my mind when I compute 13×7."

"But, they seem to have known a fact, which I was easily able to verify without even consulting a text on number theory, that any positive integer can be written in a unique way as a sum of such powers of two each counted no more than one time."

"No, I'm sorry my boy, I did not follow you through the end of this last thought. There is a *sum* of these numbers on the page?"

"That's right. You can write 13, for instance, as $1 + 4 + 8$ and not as any other sum of such powers. They would indicate this by writing a 'slash' by each power of two appearing in the sum for the first number."

After doing this, the page on which the computation is being performed looks like this:

$$\begin{array}{c} \backslash 1 \\ 2 \\ \backslash 4 \\ \backslash \underline{8} \\ 13 \end{array}$$

"Then," Mathewson continued, "one multiplies each of these powers of two by the other number in the product, the number 7 in our example, and writes those products in another column like this." He added another column so that the page now read:

$$\begin{array}{cc} \backslash 1 & 7 \\ 2 & 14 \\ \backslash 4 & 28 \\ \backslash \underline{8} & 56 \\ 13 & \end{array}$$

"Finally, one could simply *add* together the numbers in the second column that are marked by a slash: 7 + 28 + 56 = 91, which is the product we originally sought!"

"Oh, I think I see how that works as an algorithm," Professor Wilson said quietly, "but how did they arrive at *those* multiplications in the second column? Knowing, for instance, that 8 × 7 = 56 is a multiplication problem in itself."

"True, but it is a very limited sort of problem. In fact, all one had to know was how to *double* a number. Note that 14 is 7 doubled, 28 is 14 doubled and so on. So, as long as they could double the previous number, a task they could apparently do with ease since there is no indication in the papyri I looked through that they needed to do any work in computing these values."

"I see, so rather than having to memorize multiplication tables as we do now the Egyptian was able to compute any product merely by knowing how to double. That does not seem so primitive to *me*, it seems rather clever in fact!"

"Clever? You must be joking Professor Wilson. At some point in history, decimal numbers were invented, some mathematician realized the algorithm that we now use, the algorithm so simple that we now teach it to every school child, so universal that these children can use it throughout their life, whatever their station, and when this new algorithm was found, everyone saw that it was a great step forward."

The professor did not say anything, he did not need to. It was clear from his expression that he thought Mathewson, like so many others, was guilty of overestimating the brilliance of contemporary thought and underestimating the accomplishments of the past.

"Professor, you study the past looking for advanced thinking. Perhaps you are looking too hard. I cannot speak for the other sciences, I know little of them beyond what I learned in school, but I know modern mathematics and I now know something of the mathematics of ancient Egypt. There does not seem to be any way that such a backwards people could have duplicated the Theorem of Pythagoras, let alone determine the value of π. All I have seen serves to reinforce my interest in mathematics, the very reason I went into it in the first place: *progress!*"

"Progress?"

"Yes, unlike the other sciences where there are steps backwards nearly as often as forwards, mathematics always moves ahead. A theorem once proved will *always* be true and, as in this case, an algorithm that is *better* will always be better. You may reach your own conclusion, I have showed you the algorithm that they used for multiplication as you have asked."

"You certainly have, and I am very grateful. I believe I will now be better able to understand the rest of these mathematical writings. If I find that I have any further questions, may I contact you again?"

"Of course, I would enjoy that," Mathewson stood and walked to the door. Then he turned back and said "I'm sorry if I became a bit emphatic here, it's just that I would like to make clear to you how ridiculous this Egyptian algorithm for multiplication seems in light of our present understanding. Perhaps you could reasonably argue that given the limited knowledge that they had, this algorithm was brilliant *for its day*. But, what I was trying to say is that our understanding of mathematics has grown, and this method has been rendered obsolete. It is partly as a consequence of our better way of thinking that mathematical thought has brought us so far;

think of Euler, Poincaré, Gauss, and Riemann. That is mathematical progress!"

Part 3: Year 1951, Oxford University

Macintyre stood at the blackboard trying to get the attention of the class. Normally senior maths majors were not so rowdy, but today was an unusual day. Not only was this the first day they would have an opportunity to actually work with the digital computer they had been talking about all semester, but coincidentally this turned out to also be the day that the department head had chosen to attend class in order to observe Macintyre as part of the evaluation for his lectureship. All of this made Macintyre very nervous, so much so that he feared he would drop the chalk he was holding in his sweaty, quaking hands.

"Very well, let's get started then. Right. As you know, in this class we have been learning about digital computers," he said, actually addressing the man in the last row rather than the students themselves, "both to consider what these wonderful machines can do for mathematics, but also to consider what new mathematics will be needed to work with these machines. For instance, if we build a digital computer, we will want it to be able to multiply. How shall we teach it to do this? Should we teach it to multiply the same way that we do, or is there a better, *new* algorithm for multiplication that we should be using?

"It turns out," he continued though his mouth had become unbearably dry, "that the latter is true. For centuries we have been multiplying numbers in the same way, but we should not be fooled into thinking that this is the only way to do it. The one we have used, the one we learned as children, has been good enough and so there has not been a need to improve upon it ... until now. The digital computer opens a door to a new era of computation, one in which we can actually perform computations that would have been impossible to do before. This is in part because of the computer itself, but it is also due to the *program* that does the multiplication. Let me show you the clever little algorithm that the designers of computers have dreamed up. It is so simple, and requires only two basic operations, allowing it to be implemented with ease and speed on a computer."

The students in the class should, by now, have known about the binary representations of numbers, but he was not sure that Professor Mathewson would remember all he needed to know. So, Macintyre briefly reminded them of the key points. "Recall that in a computer, numbers are recorded in binary, which means that they are simply a string of 1s and 0s indicating the powers of two that add up to the number. For example, the lucky number 13 would be stored in a computer as 1101 because it is $2^3 + 2^2 + 2^0 = 8 + 4 + 1$. Now, suppose we wanted to multiply 13 by another number ..."

"Like *seven?*" suggested Professor Mathewson from the back row. This surprised Macintyre since he expected Mathewson to merely be an observer, but he was glad that the famous professor seemed to be so interested in the lecture. This was so far from Mathewson's own research in algebraic geometry that Macintyre had not really expected him to pay much attention, but Mathewson was so attentive that he seemed to have forgotten the purpose for his visit.

"And so, if we want to multiply 13 by 7, how would we go about doing this on the computer? It may not be obvious at first how to multiply two numbers when they are not written out as we normally do in decimal form, but the idea is the same in binary and turns out to be much simpler in fact. According to the distributive law of multiplication over addition, instead of taking the powers of two yielding 13 and multiplying their *sum* by 7, we can multiply each of these powers of two by 7 and then add."

To explain this, he wrote on the board

$$13 \times 7 = (1 + 2^2 + 2^3) \times 7 = 7 + 2^2 \times 7 + 2^3 \times 7$$
$$= 7 + 28 + 56 = 91:$$

Some of the students stopped taking notes and looked at Macintyre as if he were crazy.

"I know," the lecturer continued more confidently, "that this looks like trivial nonsense, but that is because you are thinking in the old-fashioned way. Think like a digital computer! Multiplication by 13 or by 7 may be something difficult, something we need to *pro-*

gram the computer to do, but multiplication by a power of two should be very easy. Why?"

There was a pause of a few seconds during which the students figured out the answer to his question, and then a pause of another minute before any of them was brave enough to raise his hand and speak up. The answer was "Because the numbers are in binary?"

"That's right! For binary numbers, multiplying by 2 is as easy as multiplying by 10 is for you and me. Do you see why this is such a nice algorithm? There is a beauty in its simplicity. Since doubling a number is now the easiest thing to do, the multiplication algorithm is built up out of nothing but doubling."

"Mr. Macintyre," called Professor Mathewson from the back of the room, "you say that this multiplication algorithm was invented by the mathematicians working on the digital computer? In trying to find an optimal multiplication algorithm, they hit upon the idea of performing a multiplication by doubling one of the numbers repeatedly and then adding together the products of this type that correspond to powers of two adding up to the first number?"

Macintyre felt as if he was being set-up, as if Professor Mathewson was trying to catch him in a mistake, but the answer to the question was clear. "Yes, that's right. That is what I said."

"And, young Macintyre, would you say that this is *progress?*"

He was not sure exactly what Professor Mathewson was getting at nor what he was expected to say. He said "Well, before this algorithm was found, they tried some others including some that mimicked our standard one. They were slow and awkward in comparison. So, yes, I guess I would say that this algorithm represented progress. Wouldn't you?"

"So, perhaps the ancient Egyptian clerics knew π, after all!" Mathewson exclaimed with a laugh.

"I'm sorry sir, but I don't think I heard you quite right. What was that again?" Macintyre said, though he was quite sure he had heard correctly.

But Mathewson did not explain the *non sequitur* to the confused class. Instead, he blinked, shook his head, stood up and left the room without saying a word.

Author's Notes

1 Unreasonable Effectiveness

The question of why mathematics is useful has been the subject of much discussion. Most famously, there is Eugene Wigner's famous article "The unreasonable effectiveness of mathematics" (*Comm. Pure & Applied Math.* 13 (1960) 1–14). But, I can't say I have ever seen mention of the explanation proposed in this story. Of course, I don't believe for a minute that it is true, but there have been strange things I didn't believe in that turned out to be true before!

In the story, Amanda's research becomes a useful mathematical model of the immune system. In fact, the human immune system remains quite mysterious today. Many attempts have been made to apply mathematical techniques to understanding it, generally only in very specific contexts. (See, for example, "A mathematical model for the investigation of the Th1 immune response to *Chlamydia trachomatis*" *Math. Biosci.* 182 (2003) by D.P. Wilson et al.) However, to the best of my knowledge, no models based on algebraic topology along the lines of those mentioned in the story have ever been developed.

I should also say something about the terms "scleroderma" and "microchimerism." These are real terms from biology where scleroderma is a serious auto-immune disease and microchimerism is an interesting phenomenon in which individual cells of one person can live indefinitely in the blood stream of another person. At the present time, there is reason to believe that these two things are related, but no conclusive evidence.

2 Murder, she conjectured

Even though Margaret Blaine and Graeme Clifton and their "theorem" are fictional, the stories of Emmy Noether and Charlotte Angas Scott are true. The sad truth is that talented female mathematicians were sometimes kept away from mathematics by a society that considered it an inappropriate activity for a woman. (It is said that the parents of Sophie Germain (1776–1831) deprived her of candles and heat at night to ensure that she did not stay up studying mathematics.) Fortunately, the situation is improving. A visit to the Website of the Association for Women in Mathematics (www.awm-math.org) will provide any interested reader both with additional historical information and resources to help women in mathematical careers today.

3 The Adventures of Topology Man

Topology is part of geometry. It is the part of geometry that has nothing to do with distances, curvature or angles. At first, one might think this is all there is to geometry, but there is more. Even ignoring all of these things, one is left with the ability to recognize whether things are *connected*, what *dimension* they have and how many *holes* there are in a given geometric object. *That* is topology.

Now, this may seem a bit weak. It is for this reason that topology is often underappreciated by the students who learn about it. Like linear algebra, it is an area that can seem overly abstract and pointless when it is first learned. However, these subjects play a fundamental role in many other areas of mathematics. It is only when one understands them in this larger context, as part of "higher" mathematical areas, that their significance can really be seen.

One of the most interesting things about topology is how often it can surprise us with counterintuitive examples. Several of these are discussed in the story: non-orientable spaces, the Klein bottle, non-Hausdorff spaces and the eversion of the sphere. The first three of these can be read about in many mathematics text books (see, for instance, the real books by Munkres or Guillemin and Pollack that the story references). The eversion of the sphere is a topological result, first attributed to Fields medalist Steven Smale, that is particularly difficult to imagine. To help people understand how a sphere can be inverted homotopically, the Geometry Center at the University of Minnesota made a film called *Outside In* that is highly recommended (even though I must admit that I still don't really understand it even after seeing it.)

In the story, two of the super-beings have topologically related powers. Topology Man has the ability to *change* the topology of spaces. The villain, Homotopy, has the complementary power of being able to alter objects in such a way that their topology is preserved.

Category Theory Girl has a much more abstract set of powers since category theory is one of the most abstract areas of mathematics. It is a field that attempts to unify such apparently different mathematical subjects as topology, group theory, geometry and linear algebra by viewing them all as examples of *categories* with *functors* and *morphisms*. At first, one might think that this is so general that it can handle *anything* ... perhaps *everything* in mathematics is just category theory in one way or another? But then, one realizes that category theory is not powerful enough to describe *itself*. (This is how Homotopy vanquishes Category Theory Girl in the story.) In fact, to really analyze category theory mathematically one needs to develop 2-Category Theory using 2- functors and 2-morphisms. And there is no reason to stop *there* either. The standard reference for category theory is the book *Category Theory for the Working Mathematician* by Saunders MacLane.

Of course, this story leaves many questions unanswered. Was Professor Wheelock *really* abducted by aliens? Are both Homotopy and Category Theory girl characters we have already met in the story or are their similarities coincidental. And, of course, how will our heroes prevail? (Unless a second issue is ever written, we may never know!)

Material quoted on p. 31 is from *Differential Topology* by Victor Guillemin and Alan Pollack (Prentice-Hall, Englewood Cliffs, NJ, 1974).

Material quoted on p. 33 is from *Topology* by James R. Munkres (Prentice-Hall, Englewood Cliffs, NJ, 1975).

4 Eye of the Beholder

I have heard many stories about the supposed mathematical abilities and interests associated with autism. It is claimed that some autistic people are capable of immediately recognizing large prime numbers, and that it is the primality of these numbers that make them beautiful to these individuals. This would be a very interesting feat, since mathematicians presently are not aware of any method for identifying prime numbers as quickly as these people supposedly can. My first reaction to such claims is skepticism, and I admit that I would not be surprised to learn that they are either exaggerated or false. However, if taken at face value, then they suggest that the human mind is capable of "running" an algorithm unknown to modern math or science for recognizing primality. Moreover, the fact that this algorithm seems to be innate rather than learned leads me to wonder whether all human brains might have this ability, even though we are not aware of it.

You can see how pursuing such a line of reasoning could lead to the thesis presented in this story, that underlying human thought is nothing other than the arithmetic properties of whole numbers. Of course, I have absolutely no evidence that this is the case, and I would not even claim to believe it. But, if one combines the fact that today's computers (that have little to do with arithmetic externally, where one sees only Flash animations and 3-dimensional WYSIWYG interfaces) certainly are just built on binary arithmetic with these stories about autistic patients having amazing mathematical abilities, it does not seem quite so far-fetched.

It should also be pointed out that the role of number theory in cryptography, and the existence of the National Security Agency, are not fiction. The NSA does employ a large number of mathematicians, and a significant part of their mission relates to the methods of cryptography that are based on number theory. In particular, the public key encryption algorithms that you probably use—without knowing it—each time you purchase something with your credit card from a Website, is an example of such a code.

There really is an aesthetic in mathematics. Truth is not sufficient. A theorem's proof will be judged for its elegance as well as for its validity. One big part of the *standard* aesthetic in mathematics seems to be designed for selecting difficult puzzles, since there is great appreciation for the simply stated but difficult to solve problems as described in the story. However, my own tastes are those that I attributed to Bev in the story.

The study of non-linear partial differential equations (NLPDEs) was of interest to many mathematicians and scientists for hundreds of years before someone realized that *some* NLPDEs were really just very simple things. These "integrable" NLPDEs then become solvable, while in the case of most NLPDEs we have to be satisfied with reasonable approximations of solutions. Moreover, they have stunning properties, such as the particle-like waves we call *solitons*. To me, this "collapse" of a difficult sort of problem into something much simpler in special cases is beautiful and interesting. However, I know those whose tastes run differently. The chaos expert Bob Devaney, for instance, once expressed his lack of interest in integrable systems to me by calling them "the dynamical systems so boring that you can solve them using algebra".

And neither of us is wrong. Beauty is in the eye of the beholder.

5 Reality Conditions

The Sumerian myth of *Gilgamesh* and his quest for immortality is a very famous story in some circles. It has a reputation for being a prototypical story, one on which countless other heroic adventures are based. Because of the opinion expressed in the G.H. Hardy quote at the beginning of the story, it seems natural to attempt to create a mathematical version of the story as well.

Even though it is not a fantasy, in *Reality Conditions*, I attempted to follow the general outline of the myth and also incorporate analogous representatives for many of the important characters in the original story. (For instance, Utanapishtim, the prototypical Noah who survives the flood intended to destroy mankind and was made into an immortal, here becomes Gil's officemate at MSRI.)

Quite a lot of things in this story are completely made up. Both Gil and his father, as well as all of the mathematicians portrayed in the story, are completely fictional. Gil's thesis work, while designed to sound plausible given what is known about quantum physics, is also just a fantasy.

However, not everything is pretend. There really is an MSRI, and I tried to make my description of it as accurate as possible. Some real mathematicians are mentioned, for instance, in Gil's debate with a physicist about the relationship between mathematics and physics. The preprint server at `arXiv.org` is real ... you can go there now to find some papers on the Hall effect, which is also real. (In fact, the braking system in my Toyota Prius makes use of it!)

And, most importantly, the goal of immortality (in the mathematical sense) is as much of a driving force to many mathematicians today as it was to Gilgamesh in his famous myth.

6 The Exception

The Goldbach Conjecture, that all even numbers greater than two are the sum of two primes, is one of the most famous open problems in mathematics. This is not because it is especially important but rather because it is relatively simple to state and yet apparently very hard to prove. Fermat's Last Theorem held a similar status until it was proved in the 1990's by Andrew Wiles and Richard Taylor.

As I write this, it is still completely unknown whether it is true. Nobody knows how to show it is always true and nobody knows a case in which it does not work. Vector bundles are real mathematical objects and it is *conceivably* possible that someone could prove Goldbach's conjecture by showing that there is always a vector bundle of rank $2n$ on a torus that is the direct sum of two vector bundles of prime rank. However, I have no reason to think that such a proof exists; this is pure fiction.

7 Pop Quiz

Of course, this story is completely fictional and some might even say that it is 'far fetched.' But, it has more of a basis in reality than you might think! As a professor of mathematics, I quite frequently get e-mail from college, high school and middle school students from around the US and around the world, asking for help with their homework assignments. It is sometimes rather specific (e.g., "What is the formula for the 2-soliton solution to the KdV equation?"). One simply said: "I have to write a paper on surfaces. Can you give me information?" A local student here in Charleston wrote to the college and said that he had a report due in two weeks and had done nothing yet. He was hoping that for his report, one of the professors here would go to his class and make a presentation! The point is this: having been given access to the internet, a method for communicating almost instantaneously with experts in any part of the world, these students prefer to ask strangers for help rather than do the work themselves. And, with the huge number of people available to them on the internet, I wouldn't be surprised if they can generally find at least one person willing to do their assignment for them.

Some of the mathematics presented in this story is real, in particular, the stuff about projective coordinates, grassmannians and grassmannian duality. (I left out a little detail about the grassmannian duality: we need a little bit of extra structure, like an inner-product on the underlying vector space, for the duality to exist.) However, the stuff Sarah figures out in answering the third question is completely made up, and probably makes no sense if you think about it carefully.

8 The Math Code

A friend of mine who is not a mathematician once complained that mathematical notation seems to have been constructed with the intention of making our writings obscure. He thought that we were protecting our ideas by making them incomprehensible to "outsiders". This is not at all the case: the symbols were not devised as a code but as an *argot*. The creation of mathematical symbols is merely an attempt to communicate our abstract mathematical ideas to one other. Still, as illustrated in this story, that does not mean that we cannot also take advantage of them as a cipher when the need arises!

9 Monster

Yes, Virgina, there really is a Monster! Like my description of group theory in this story, I have tried to make my description of the Monster group as accurate as possible in this context. For more information, consider a professional resource such as *Sporadic Groups* by Michael Aschbacher.

I may have overemphasized the significance of the need for a geometric interpretation of the Monster group. It is not *completely* geometry free. (In fact, it was originally defined by Robert Griess as a group of rotations in a 196,883-dimensional space.) Moreover, many algebraists probably are not at all uncomfortable with ignoring any geometric interpretations it may have. However, I have heard an expert remark that he feels a *natural* geometric interpretation is lacking and that the discovery of one would allow us to better understand some of the Monster's remaining mysteries.

There are puns scattered throughout this story (including most of the characters' names, and the Ogden-Nash-Hamilton equation that is evocative of both John Nash's famous Nash-Hamilton theorem and the poet Ogden Nash), and ideas for mathematical "inventions" that are probably not really possible. However, one of the less believable parts of the story—a student taking a make-up test unsupervised in the professor's office copying answers from another test left in plain sight on the desk—really did happen in a class that I TA'd as a graduate student!

10 The Corollary

The pressure to publish can be quite a motivator in any area of academia, and as a result there are some "tricks" people use to increase the number of their publications. However, I have never come across an example of "cheating" quite as blatant as I present in this story.

For instance, I have heard people in supersymmetric integrable systems brag at conferences that their area of research is great because they get to publish each result twice: once in a supersymmetric setting and once in a standard commutative setting. Of course, the fact that they were willing to announce this publicly suggests that it is not a problem like the secret methods of the protagonist in this story. Similarly, some mathematicians are able to publish their articles twice because they publish regularly in English and in another language. (Actually, even though I do not speak Russian, I do have an article that was published both in English and in Russian.) Moreover, although I have never heard of an author intentionally leaving out an important lemma as the mathematician in the story does, it would be unreasonable to expect authors to avoid sending their journal submissions to editors they consider friendly.

In this story, I create a hypothetical (unethical) mathematician who makes use of all of these tricks and *more* to increase the length of his CV and I point out that this only works as long as one is doing research that does not get a lot of attention.

11 Maxwell's Equations

The main historical facts presented in this story are, to the best of my knowledge, true. In particular, James Clerk Maxwell was the first to unite the theories of electricity and magnetism by writing a wave equation, and he proposed that what we call light is nothing other than a special case of these electro-magnetic waves. His *Treatise on Electricity and Magnetism*, that contains the famous Maxwell equations, was finally published in 1873. Maxwell died in 1879, but his equations were tremendously important to 20th century physics. As he had predicted, it was possible to produce and receive these invisible waves as electrical currents. By the late 1880's, Heinrich Hertz was able to send such a signal *several feet*. In 1899, Marconi shocked the world by sending "radio signals" across the English Channel.

In fact, one could argue that radio, television, radar, microwave ovens, lasers, and even Einstein's theory of relativity all grew, in one way or another, out of Maxwell's formulas for electro-magnetic waves. However, in saying this I enter highly contentious territory. History, as ever, is not as clear and simple as a story. Certainly there was some indication prior to Maxwell's equations that electricity and magnetism were related, and although there is no denying the importance of the equations, whether much of the theory would have been discovered soon in a less mathematical way is the subject of debate today.

This is all irrelevant to the point of the story, however. My goal here is not to relate a piece of history—remember that the presentation here is entirely fictional—but to convey to the reader the wonderful feeling that one has when the equations all fall into place and everything starts to make sense. When that happens, even in the most trivial of circumstances, it can feel as if the universe has revealed one of its secrets to you. No matter what the true circumstances surrounding the discovery of Maxwell's equations, this must be how he felt.

(I should also mention that the Voyager space probes confirmed his description of Saturn's rings.)

12 Another New Math

There is a traditional *order* in which mathematics is taught. For instance, the first non-commutative structure that math students come across in the traditional curriculum is the cross product of two vectors which shows up in multivariable calculus and has the property

$$\vec{v} \times \vec{w} = -\vec{w} \times \vec{v}.$$

Matrix multiplication is another non-commutative algebraic structure that is encountered by undergraduate students majoring in physics, mathematics or computer science. Most students never see any examples of non-commutativity. Nevertheless, non-commutativity is quite normal. Among all possible algebraic structures studied by algebraists, commutativity is a rare property. Moreover, the real world presents many examples of non-commutativity (both simple and deep). The famous example of opening a window and sticking your head out (leading to very different results depending on the order you choose to do them in) is a non-mathematical example. However, the solution of a "Rubik's Cube" is an easily understood case in which non-commutative algebraic structures really are useful. (See, for instance, David Joyner's book *Adventures in Group Theory*.) More abstractly, but perhaps more importantly, the development of quantum theory in the early 20th century revealed that measurements are non-commutative! (You'll see many philosophical and religious interpretations of Heisenberg's uncertainty principle in popular physics books, but from a mathematical point of view it merely says that when computing the product of the position and momentum of a particle, the order matters. In other words, these things are "operators" and not numbers.)

Similarly, nonlinearity is a mathematical subject that is generally hidden or avoided. In the 19th century, little was known about the world of nonlinearity. However, in the 20th century we came to know about its two extremes: chaos and integrability. We now understand nonlinearity much better, and recognize that objects from chaos theory and integrability (fractals and solitons respectively) may play an important role in our further understanding of the world around us.

Although I do not really believe that learning about these mathematical subjects earlier would allow us to teleport our bodies from room to room, I do wonder whether we are jeopardizing our ability to understand reality intuitively when we continue to focus almost exclusively on commutativity and linearity despite the recognition that the real world does not always satisfy these simplifying assumptions.

13 The Center of the Universe

There are two real mathematical facts that are relevant here. First of all it *is* true that almost every number contains every finite sequence in its decimal expansion. Though one should remember that this is a sort of *statistical* comment. Notice that the numbers we deal with most frequently, like integers and rational numbers, do *not* have this property. Moreover, it is very difficult to check whether a given irrational number has this property. For instance, I don't think it is known whether the number π has the property or not! For details see Theorem 146 in the book *An Introduction to the Theory of Numbers* by G.H. Hardy and E.M. Wright.

Also, there really is an algorithm for computing the nth digit of the number π without knowing any of the other digits and it really only works in base two or base sixteen, as far as anyone knows. The original paper by Bailey, Borwein and Plouffe in which this was proved is presently available on the internet at the URL:

www.lacim.uqam.ca/~plouffe/articles/BaileyBorweinPlouffe.pdf

in case you are interested in seeing how it works or trying it for yourself.

However, there is one piece of background I should mention here that is *fictional*. The novel *Contact* by Carl Sagan has as its conclusion a scene in which the protagonist finds a 'hidden message' in the number π. In particular, she finds a string of digits in the base 17 expansion that are all zeroes and ones and that draw a perfect image of a circle when displayed as pixels on a computer monitor. There are lots of questions that this discovery brings up that are *not* addressed by Sagan. So, in writing this story, I'm just trying to bring these questions to the forefront. Namely, given Hardy's result, what would such a discovery imply? Should we conclude that the universe was carefully constructed so that this natural constant relating the diameter and circumference of a circle has a picture of a circle in it? Is it just a coincidence? Is it a near *certainty* that the expansion of the number π actually does contain such a number according to Hardy, and if so wouldn't it still be interesting if the sequence was found?

Many people I have discussed this with find it disturbing to think that an irrational number contains every possible finite sequence. It would certainly have some scary sounding consequences. For example, it means that any letter that you are going to write in the future already appears, word for word, as a sequence in that number. But, if you think about it in another way, it becomes clear that this is not so strange. Imagine (as Borges and Lasswitz have done in their stories) a library of books where each book just contains a random string of one million letters, punctuation marks or spaces. Notice that, although there are a *lot* of sequences like that, there are only a finite number of them, so you could have all of them in a library with only finitely many books.

Some of these books would be silly (like all 'a's or just nonsense) but among them would be a copy of Twain's *Pudd'nhead Wilson*, *Gone with the Wind* and the novel that is going to be a best seller next year but has not yet been written. Now, given that you can get all of that in a finite number of books, it is not too surprising that you can also get it in the *infinite* sequence that is the decimal expansion of an irrational number!

One last "bit" I should explain is the reference to Shannon and Tukey. Pioneering statistician John Tukey is credited with coining the term "bit" in 1949 to mean "binary digit" as it is presently used in computer science. Claude Shannon, the mathematical engineer who founded the science of information theory, has a similar notion of a bit as a unit of information. However, there is a subtle distinction in that a computer file that takes up n bits on a computer's hard drive (the Tukey sense of bit) may actually have fewer than n bits in the Shannon sense if it could be compressed to a smaller size without losing any information.

14 the object

Much of the mathematics discussed in this story is real. In particular, Euler did show that for any polyhedron without holes in it, the number of vertices minus the number of edges plus the number of faces is always 2. This fact was then used in the proof that there are only 5 regular polyhedra. It was also further generalized to the case of more complicated topologies (that is, polyhedra with holes in them) and even more abstract areas of mathematics (where the Euler characteristic is a rather general geometric invariant). Two informative and entertaining presentations of this concept can be found in mathematical fiction. An elementary and amusing presentation is given in Hans Enzenberger's *Number Devil*, and a more ambitious approach is taken in Greg Egan's *Diaspora*.

Similarly, the discussion of solitons, instantons and geometry is also intended to be as realistic as possible (with the obvious exception that so far none of these mathematical topics has resulted in a bridge to another universe through which small, shadow-like creatures come to devour us!) For instance, Euler's proof that there is no regular polyhedron of the type discussed in the story was dependent upon the particular geometry of flat Euclidean space. It is conceivable (although, I admit I have not actually verified this) that an alternative geometry would allow for their existence. That geometry could have some interesting features, just as instantons really are examples of interesting geometries that are surprisingly like 4-dimensional Euclidean space ... but not exactly. (One of the most amazing results of 20th century mathematics is the discovery that it is only for four-dimensional space that such distortions can exist.) These instantons are a form of geometric soliton (a localized wave phenomenon.) As described in the story, topological solitons are studied in mathematical physics as particles in a quantum field, but can also be interpreted as a localized connection between two different vacua. When the connection between the same vacua occurs but with the opposite orientation, one has what is known as an *anti*-soliton, an object that will annihilate the corresponding soliton upon collision like matter and anti-matter.

15 The Legend of Howard Thrush

The "tall tale" is a tradition of American folk literature and an important part of American culture. The stories of the great heroes of the untamed West and the way these stories explain the features of our landscape (e.g., Pecos Bill carving out the Rio Grande to bring people water during a drought) were more than just entertainment to the people on the frontiers.

Mathematical research is a sort of frontier as well. Certainly, mathematics also has its legends and heroes: little Carl Gauss outsmarting his teacher by being able to compute a sum with many terms in a clever way, Isaac Newton and his apple, Sonya Kovalevskaya convincing the great mathematician Karl Weierstrass that perhaps a woman *could* do mathematics after all. I am not certain to what extent these stories are just legends and to what extent they are true, but they all fall a bit short of the hyperbolic achievements of Paul Bunyan or Mike Fink.

In any case, if mathematics is ever in need of a real tall tale hero, then it might now have one in Howard Thrush. He certainly is completely fictional; it is difficult for me to think of many true things to discuss in the story.

I should mention that the "sharp inequality" he proves to escape from the rustlers is a famous theorem of Bieberbach. When mathematicians say that it is sharp, we mean that although the 2 in it could be replaced by a larger number leaving the statement true, it could *not* be replaced by any smaller number. Also, his escape from the rustlers was aided by an understanding of the Poincaré formula that relates the numbers of vertices, edges and faces of a triangulation to the genus of the surface.

16 Progress

The characters in this story are all completely fictitious. However, the ironic history of this algorithm is, to the best of my knowledge, entirely true. The mathematicians of ancient Egypt did use the algorithm described in this story, as I have learned from the wonderful book *Mathematics in the Time of the Pharoahs* by Richard J. Gillings. From the same source I have learned that mathematicians and historians of the early 20th century found the algorithm to be primitive, and referred to it as one limitation in the scientific growth of the ancient Egyptian culture. Finally, it is also true that the multiplication algorithm used by early digital computers was essentially this same algorithm and that it is this same algorithm, now in the form of hardware rather than software—built into the components of semi-conductor chips—that performs the multiplication for nearly every book-keeper, scientist and mathematician. (Thanks to Steve Newman for his input on how computers really do multiply.)

About the Author

After receiving his PhD in mathematics from Boston University in 1995, Alex Kasman held postdoctoral positions at the University of Georgia in Athens, The Centre de Recherches Mathématiques in Montréal and the Mathematical Sciences Research Institute in Berkeley. Much of his research involves the application of techniques from algebraic geometry to problems in analysis and mathematical physics. In particular, his work addresses the questions of commutativity and bispectrality of differential operators and the dynamics of waves and particles. In addition, he has recently published a paper in mathematical biology, written jointly with his wife, that has become the basis of a patent application filed by an international pharmaceutical company. This research has been published in prestigious journals in mathematics, physics and biology and presented at the Newton Institute of Mathematical Sciences in Cambridge. Alex lives with his wife and daughter in Mount Pleasant, South Carolina where he is an Associate Professor of mathematics at the College of Charleston.